符号中国 SIGNS OF CHINA

中国古代科学

SCIENCE IN ANCIENT CHINA

"符号中国"编写组 ◎ 编著

中央民族大学出版社
China Minzu University Press

图书在版编目(CIP)数据

中国古代科学：汉文、英文 /"符号中国"编写组编著. —北京：中央民族大学出版社, 2024.3
（符号中国）
ISBN 978-7-5660-2280-6

Ⅰ.①中… Ⅱ.①符… Ⅲ.①自然科学史—中国—古代—汉、英 Ⅳ.①N092

中国国家版本馆CIP数据核字（2024）第017052号

符号中国：中国古代科学 SCIENCE IN ANCIENT CHINA

编　　著	"符号中国"编写组
策划编辑	沙　平
责任编辑	杨爱新
英文指导	李瑞清
英文编辑	邱　械
美术编辑	曹　娜　郑亚超　洪　涛
出版发行	中央民族大学出版社
	北京市海淀区中关村南大街27号　　邮编：100081
	电话：（010）68472815（发行部）　传真：（010）68933757（发行部）
	（010）68932218（总编室）　　　　（010）68932447（办公室）
经 销 者	全国各地新华书店
印 刷 厂	北京兴星伟业印刷有限公司
开　　本	787 mm×1092 mm　1/16　印张：8.125
字　　数	105千字
版　　次	2024年3月第1版　2024年3月第1次印刷
书　　号	ISBN 978-7-5660-2280-6
定　　价	58.00元

版权所有　侵权必究

"符号中国"丛书编委会

唐兰东　巴哈提　杨国华　孟靖朝　赵秀琴

本册编写者

王　慧

前言 Preface

　　科学，是反映现实世界各种现象的本质和规律的知识体系。中国是世界四大文明古国之一，拥有上下五千年的文化历史。自原始社会时期开始，中国古代科学就已经起步。经过漫长的历史发展，中国古人在天文学、地理学、数学、医药学和农学等

Science is the knowledge system that reflects the nature and objective laws of all sorts of phenomena in the real world. As one of the Four Great Ancient Civilizations, China has a history which can stretch back over five thousand years. Since the primitive society, science had started sprouting in ancient China. After a long history

方面取得了许多科技成果，并创造了许多个世界第一。这其中既有对圆周率值的计算，对天体的精确观测，也有给全世界带来深远影响的造纸术、印刷术、火药、指南针等。

本书从天文学、地理学、数学、医药学、农学五大科学领域，系统地为读者介绍中国古代科学方面的知识。

of development, numerous outstanding scientific accomplishments were achieved and inherited regarding to the areas of astronomy, geography, mathematics, medicine, agronomy, etc. And many of the world's No.1 inventions or improvements were created, including the calculation of π, the accurate astronomical observation, as well as the Four Great Inventions (papermaking, printing, gunpowder and compass) that exerted great influence to the whole world.

This book introduces the science in ancient China systematically from five major scientific fields of astronomy, geography, mathematics, medicine and agronomy.

目录 Contents

天文学
Astronomy 001

宇宙学说
Cosmological Theories 002

天文观测
Astronomical Observations 008

天文历法
Astronomical Calendars 015

地理学
Geography 025

古代地理学概述
Overview of Ancient Geography 026

古代地理学著作
Ancient Works on Geography 034

数学
Mathematics 049

算筹与筹算
Counting Rods and Rod Arithmetic 050

算盘与珠算
Abacus and Abacus Arithmetic....................... 053

古代数学著作
Ancient Works on Mathematics 058

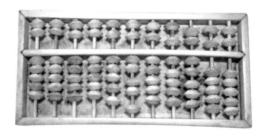

医药学
Medicine... 065

中医理论
Theories of Traditional Chinese Medicine 066

中药学
Science of Chinese Pharmacology 072

古代医学典籍
Ancient Canons of Traditional
Chinese Medicine 077

农学
Agronomy 091

农业生产与二十四节气
Agricultural Production and 24 Solar Terms 092

农学家与著作
Agriculturists and Works on Agronomy 103

水利灌溉工程
Water Conservation and Irrigation Projects 112

天文学
Astronomy

  中国是世界上天文学起步最早、发展最快的国家之一。早在6500多年前，中国古人就已经掌握了二十八宿、北斗七星及日月变化的规律。在宇宙学说、天文观测、天文历法和仪器制作四个方面，中国古代的天文学都取得了重大突破。

China is one of the countries with an earliest start on the study of astronomy and the greatest development. As early as over 6,500 years ago, ancient Chinese had mastered the regular changing pattern of the 28 Chinese Lunar Mansions, the Big Dipper and the variations of the sun and the moon. The astronomy in ancient China has achieved major breakthroughs in four areas of cosmological theory, astronomical observation, astronomical calendar and instrument production.

> 宇宙学说

"盖天说"是中国古代最早的一种宇宙学说，即天是圆的，像一把张开的大伞覆盖在地上，地是

> Cosmological Theories

The Theory of Canopy Heaven is the earliest cosmological theory in ancient China. According to this theory, the heaven is round, like an open canopy covering the earth which has a square shape like a chessboard, while the sun, the moon and the stars are passing back and forth like reptiles on the canopy. Thus, the theory is also known as Theory of Round Heaven and Square Earth. The theory believes that the sun, the moon and the stars do not really come in and out, but just out of our sight when they leave far away and into our sight when they move close. The theory reflects a

• 张衡像
Portrait of Astronomer Zhang Heng

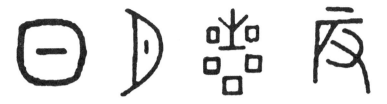

- 日、月、星、辰（甲骨文）
  甲骨文是迄今为止中国发现的年代最早的文字，出现在殷商时期。
  The Sun, the Moon and the Stars (Oracle)
  Oracle is so far the earliest Chinese characters found in China, which emerged in the Shang Dynasty (1600 B.C.-1046 B.C.).

方的，像一个棋盘，日月星辰则像爬虫一样在天空来回经过，因此又称"天圆地方说"。"盖天说"认为，日月星辰的出没，并非真正的出没，而只是离远了就看不见，离得近就看见了。它反映了人们认识宇宙结构的一个阶段，是中国古人在描述天体的视运动方面的最初尝试。

由于"盖天说"只是古人根据对自然的观察所得出的浅显理论，并不能解释诸如日月星辰东升西落的真正原因，因此在战国时期（前475—前221）"浑天说"逐渐产生。东汉时期的天文学家张衡（78—139）在其著作《浑天仪注》中明确提出"浑天说"这一更

stage when human cognize the structure of the universe, and is the first attempt of ancient Chinese to describe the apparent motion of celestial bodies.

Derived by the ancients on the basis of the observation of nature, the Theory of Canopy Heaven was too plain to explain why the sun, the moon and the stars rise from the east and sink in the west. Thus, in the Warring States Period, the Theory of Sphere Heaven gradually emerged. In the Eastern Han Dynasty (25-220), astronomer Zhang Heng (78-139) explicitly put forward this more advanced cosmological theory in his book *Notes on Armillary Sphere*: "The sphere heaven is like an egg. Celestial bodies are round like beads while the earth sits lonely within the heaven just

加先进的宇宙理论："浑天如鸡子，天体圆如弹丸，地如鸡子中黄，孤居于天内，天大而地小。"

- 北京古观象台

北京古观象台始建于元代，原名"司天台"，清代时，更名为"观象台"。北京古观象台以仪器精美、观测准确而著称。

Ancient Observatory in Beijing

The ancient observatory in Beijing was built in the Yuan Dynasty (1206-1368). It was originally called *Sitian Tai*, which was changed to *Guanxiang Tai* (meaning observatory in Chinese) in the Qing Dynasty (1616-1911). The observatory is famous for its exquisite instruments and precise observations.

as the yolk of egg; the heaven is larger than the earth." Zhang Heng believes that the earth is like the yolk of egg while the heaven looks as if it is wrapped with eggshell. The heaven is much larger than the earth. Half of the heaven is covering above the earth and another half under the earth. That's why half of the 28 Chinese Lunar Mansions appear when another half of them disappear. Zhang Heng also believes that the universe is infinite and that there are many things about it human do not know yet. He accordingly made

他认为地就像一个鸡蛋黄，而天则好似一个鸡蛋壳包裹着它，天很大而地很小，且天一半覆盖着地上，一半覆盖着地下，所以天上的二十八星宿才会时隐时现。张衡还认为宇宙是无极限的，有许多规律是人类还不了解的。他据此制作了观测天象的仪器浑仪，根据浑仪所观测的天象而制定的历法十分精确。张衡还创制了可以演示宇宙天体运动的浑象，进一步证明了"浑天说"的可信度。到了唐代，"浑天说"逐渐代替"盖天说"成为古代天文领域的指导理论。

除了"盖天说""浑天说"，还有另外一个令世人瞩目的宇宙理论学说——"宣夜说"。在战国时期，"宣夜说"就已出现。"宣夜说"认为无限的宇宙中充满着气体，所有天体都在气体中漂浮运动，日月星辰的运动规律由其本身的特性所决定，并非被固定在某个球体坚硬的外壳轨道上。

"盖天说""浑天说""宣夜说"等宇宙学说较之欧洲的早期宇宙学说，例如古希腊学者欧多克斯

an astronomical observation instrument, the Armillary Sphere. By using the Armillary Sphere to observe the celestial phenomena, he developed a very precise calendar. He also made the Celestial Globe to demonstrate the movement of cosmic objects, which further proved the credibility of the Theory of Sphere Heaven. By the Tang Dynasty (618-907), the Theory of Sphere Heaven gradually took the place of the Theory of Canopy Heaven and became the guide theory in ancient astronomy.

In addition to the Theory of Canopy Heaven and the Theory of Sphere Heaven, another cosmological theory also wins the attention of the world: the Theory of Infinite Space, which has emerged since the Warring States Period (475 B.C.-221 B.C.). This theory believes that the infinite universe is filled with gas, and all celestial bodies are floating and moving in the gas. The sun, the moon and the stars have the laws of motion determined by their own characteristics, so they are not fixed on orbits of a hard sphere's shell.

All the above cosmological theories appeared earlier than those developed in Europe, such as the Geocentric Theory put forward by ancient Greek scholar

提出的"地心说"、波兰天文学家哥白尼在1543年提出的"日心说"出现的时间都要早。这三大宇宙学说奠定了中国天文学的理论基础。

Eudoxus and the Heliocentric Theory put forward by Polish astronomer Copernicus in 1543. The three cosmological theories have laid the theoretical foundation for Chinese astronomy.

盘古开天辟地

在中国古代神话传说中,盘古是生活在混沌之中的神。他不能忍受这样的生活,于是就用一把神斧将混沌劈开了。混沌分成了天与地,盘古将自己的身躯横在中间,以防止天与地重新合起来。天不断上升,地不断下沉。随着天地的不断移动,盘古的身躯也慢慢拉长,终于有一天他再也支撑不住而倒下了。他的左眼变成了太阳,右眼变成了月亮,眼泪变成了星星,汗珠变成湖泊,血液变成了江河,筋脉变成了道路,毛发变成了草原和森林,呼出的气体变成了清风和云雾,发出的声音变成了雷鸣。人们将盘古尊为开天辟地的创世之神。

盘古开天辟地的神话,是中国古人对宇宙的最初猜想,表明原始社会的人们对宇宙还没有一个清楚的认识。

Pangu Created the Heaven and the Earth

According to ancient Chinese mythologies, Pangu was a god lived in the chaos. He could not stand this kind of life, so he used an holy axe to split off the chaos, dividing it into the heaven and the earth. To prevent the heaven and the earth from reconsolidating, Pangu erected his body between them. As a result, the heaven continued to rise and the earth continued to sink. With the constant movement of the heaven and the earth, Pangu's body was stretched gradually, and then one day he could no longer stand anymore and fell down finally. His left eye turned into the sun, right eye turned into the moon, and tears into stars, sweats into lakes, bloods into rivers, tendons and vessels into roads, hair into grasslands and forests, exhaled gas into breeze and clouds, and his voice into thunders. Therefore, people respect Pangu as the Creator of the heaven and the earth.

The myth of Pangu's Creating the Heaven and the Earth indicates that people in the primitive society do not have a clear understanding of the universe, and the myth is just their primary imagination of the universe.

- 盘古开天辟地
  Pangu Created the Heaven and the Earth

> 天文观测

　　人类早期对于日月星辰的认知十分有限，常常认为世界万物的存在都是神的旨意。中国古人很早就开始"夜观星象"，用其现象和变化来对照国家的命运兴衰。每个朝代的皇帝都自诩是"天子"，即上天的儿子，为了彰显自己的尊贵，他们设立专门的天文机构用来占卜上天的旨意。秦代、汉代设"太史令"，唐代设"太史局"（后改称"司天台"），明清时期则设"钦天监"。历史上许多著名的天文学家本身也是占星师。

　　中国古代在哈雷彗星、太阳黑子和流星雨的天文观测上，所取得的成就举世闻名。西汉时期的古籍《淮南子·兵略训》中记载："武王伐纣，东面而迎岁，至汜而水，

> Astronomical Observations

Early human's knowledge about the sun, the moon and the stars was very limited, and they often thought that everything in the world was the will of God. At a very early stage, the ancient Chinese started the observation of astrology at night, as a means of divining the fate of the country. Emperors of each dynasty praised themselves as the Son of Heaven, so as to highlight their dignity. Special astronomical institutions were established to divine the will of Heaven. In the Qin Dynasty (221 B.C.-206 B.C.) and Han Dynasty (206 B.C.-220 A.D.), the government applied the position of imperial astronomer (*Taishi Ling*, ancient position in charge of recording history, managing books, calendric system and ritual ceremony); then in the Tang Dynasty (618-907) it applied the

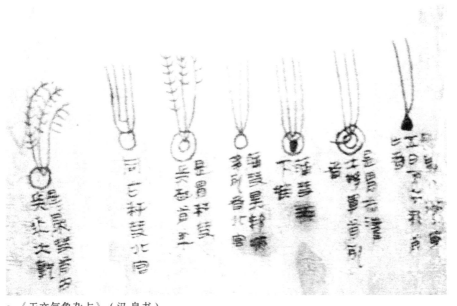

- **《天文气象杂占》(汉 帛书)**

  这幅帛书上记载了天文观测到的彗星图像，绘制有不同形状的彗尾，是世界上最古老的彗星图。

  Silk Manuscript: *Miscellaneous Divinations on Astronomy and Meteorology* (Han Dynasty, 206 B.C.-220 A.D.)

  This silk manuscript recorded the observed comet images with different tails of comets, which are the world's most ancient comet images.

至共头而坠，彗星出，而授殷人其柄。"周武王领兵讨伐施行暴政的商纣王期间，天空中出现了彗星，自西向东划过天空。这颗彗星就是哈雷彗星。据此，历史学家推测出武王伐纣之年应该在公元前1057年。在春秋时期的史书典籍《春秋》中记载，鲁文公十四年（前

administrative office of Astronomical Bureau (*Taishi Ju*, later called Administration of Heavenly Observatory, *Sitian Tai*); and in the Ming and Qing dynasties (1368-1911), it set the office of Imperial Astronomical Bureau (*Qintian Jian*). Many famous astronomers in history were also astrologers.

The ancient Chinese have made

613年）"秋七月，有星孛入于北斗"，这是有关哈雷彗星的最早有确切纪年的历史记录。从公元前1057年至公元1986年，哈雷彗星共回归地球41次，中国古人不但记录了最早的3次，而且对公元前240年至公元1986年的30次回归都有详细记录。

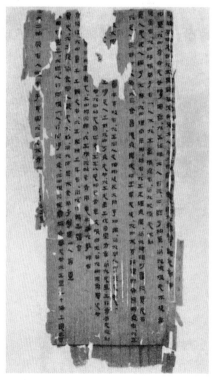

• 《易经》（西汉 帛书）（图片提供：FOTOE）
Silkbook *The Book of Changes* (Western Han Dynasty, 206 B.C.-25 A.D.)

world-famous achievements in the astronomical observation of Halley's Comet, sunspots and meteor showers. *Huai Nan Zi: Bing Lue Xun*, an ancient book of the Western Han Dynasty (206 B.C.-25 A.D.), recorded that: "King Wu was sending a punitive expedition against King Zhou, and then the Jupiter appeared in the east (meaning bad things will happen). Floods rushed down when the troops arrived at Si River, and landslide struck them when the troops arrived at Gongtou Mountain. Then, a comet appeared in the sky, whose tail was pointing to the territory of King Zhou (meaning the God was in King Zhou's side)." In 1057 B.C. when the King Wu of Zhou Dynasty (1046 B.C.-256 B.C.) was sending a punitive expedition against the King Zhou of Shang Dynasty (1600 B.C.-1046 B.C.), the sky appeared a comet, which crossed over the sky from the west to east. That was the Halley's Comet. According to *The Spring and Autumn Annals*, a history book of the Spring and Autumn Period (770 B.C.-476 B.C.), "in the autumn of 613 B.C., a comet flew into the Big Dipper", which is the earliest historical record about comet. From 1057 B.C. to 1986 A.D. Halley's

除了对彗星有早期的观测外，中国古人对太阳黑子的观测也由来已久，而且在望远镜没有发明之前，都是用肉眼进行观测、记录的。《汉书·五行志》中记载，公元前28年"有黑气大如钱，居日中央"。《易经》中也有"日中见斗"的记载，"斗"即太阳黑子。在其他史书中还常有"日中有黑子""日中有黑气"等相关记载。

如今，流星雨的观测与记载已十分常见，出现时间也可准确预报。但是，在战国时期的史书《竹书纪年》中就曾记录过公元前16世纪商朝出现的一次流星雨："夏帝癸十五年，夜中星陨如雨。"这是世界上最早的关于流星雨的资料。史书《左传》中还有关于鲁庄公七年（前687年）流星雨的记载："夏四月辛卯夜，恒星不见，夜中星陨如雨。"这是世界上天琴座流星雨的最早记录。

战国时期，齐国的甘德和魏国的石申是两位最为突出的天文学家，他们都对恒星进行了长期细致的观测，并各自建立了不同的全天恒星区划命名系统。甘德

Comet returned to the Earth 41 times, and the ancient Chinese not only made records of the first three times, but also made them in detail for the 30 returns from 240 B.C. to 1986 A.D.

Apart from the early observation of the comet, the ancient Chinese also has a long history in the observation of sunspots. And before the telescope was invented, such observations and records were finished with the naked eyes. According to *Han Shu: Wu Xing Zhi*, in 28 B.C., "There was black gas as big as a coin in the middle of the sun." *The Book of Changes* also has a record that "there was sunspot in the middle of the sun", and in other history books, relevant records can also be found, such as, "there was black mole in the middle of the sun", and "there was black gas in the middle of the sun".

Today, the observations and records of meteor shower are very common, and the forecasts about the time of emergence are also of high accuracy. However, a meteor shower appeared during the Shang Dynasty in the 16th century B.C. has been recorded in a history book of Warring States Period (475 B.C.-221 B.C.) *Bamboo Annals*: "In the fifteen year

● 马王堆一号汉墓"T"形帛画

这是1972年在湖南长沙马王堆一号汉墓中出土的帛画，上方画着一轮红日，中间画有一只乌鸦，似象征着太阳黑子的出现。

T-shaped Silk Painting

On the silk painting unearthed from the No.1 Tomb of Mawangdui in Changsha of Hunan Province in 1972, there is a red sun with a crow in the middle of it, which may symbolize the emergence of sunspots.

of the Reign of Emperor Gui in the Xia Dynasty (approx. 2070 B.C.-1600 B.C.), the meteorites fell down like rain at night." This is the world's earliest record about meteor shower. Also, the history book *Zuo Zhuan* has a record about the meteor shower in 687 B.C.: "One night in summer, fixed stars disappeared and meteorites fell down like rain at night." This is the world's earliest record about the meteorites from Lyra.

During the Warring States Period (475 B.C.-221 B.C.), Gan De from the State of Qi and Shi Shen from the State of Wei were the most prominent astronomers, who made long-term and careful observations on fixed stars in the sky, and respectively established different all-day compartmenting and naming systems for fixed stars. Gan De recorded the moving situations and laws of the five planets of Venus, Jupiter, Mars, Mercury and Saturn, and he also discovered the retrograde motion of Mars and Venus. He pointed out that: "The moving path of anterograde-retrograde-anterograde motion, is in a form of the Chinese character ' 巳 '." He pioneered the concept of planetary synodic period (celestial bodies' rotation and revolution

记录了金星、木星、火星、水星、土星五颗行星的运行情况及规律,还发现了火星、金星的逆行现象。他指出"去而复还为勾""再勾为巳",意思是行星从顺行到逆行、再到顺行的运动轨迹为"巳"字形。他首创了行星会合周期(相对于太阳,天体的自转和公转周期)概念,并测得木星、金星和水星的会合周期值分别为400日(实为398.9日)、587.25日(实为583.9日)和136日(实为115.9日)。

石申对金星、木星、水星、火星、土星等行星的运行进行了细致的观测和记录。石申在观察木星时发现"若有小赤星附于其侧",这颗小赤星就是其卫星木卫二。直到1610年,意大利科学家伽利略才用天文望远镜观测到这颗卫星。

甘德与石申的著作被后人结合在一起,命名为《甘石星经》,这是世界上最早的天文学著作。书中共记录了800多颗恒星的名字,并划分为若干星官,同时提及日食、月食是天体相互掩食的现象。现在月球上的一座环形山便是以石申的

periods relative to the sun). He measured the synodic period of Jupiter as 400 days (the correct value is 398.9 days), Venus as 587.25 days (the correct value is 583.9 days), and Mercury as 136 days (the correct value is 115.9 days).

Shi Shen made careful observations and records on the motions of Venus, Jupiter, Mercury, Mars, Saturn and other planets. When Shi Shen was observing Jupiter, he found that: "It looks as if there is a small red planet beside Jupiter." That small red planet is Europa, which was not discovered by other astronomers outside China until 1610, the Italian scientist Galileo found it with the assistance of telescope.

The works of Gan De and Shi Shen later was combined into a book named *Gan Shi Xing Jing*, which is the world's oldest astronomical works. The book has the records of over 800 stars' names which are divided into asterism. It also mentions that solar and lunar eclipses are the phenomenon of celestial bodies' mutual occultation. A crater on the moon now is named after Shi Shen.

The list of stars in *Gan Shi Xing Jing* was finished about 200 years earlier than the first list of stars in Europe,

● 日食（图片提供：微图）
Solar Eclipse

名字命名的。

《甘石星经》中的恒星表比希腊天文学家伊巴谷在公元前2世纪测编的欧洲第一个恒星表早了约200年。后世许多天文学家在测量日、月、行星的位置和运动时，都用到了《甘石星经》中的数据。

which was measured and compiled by Greek astronomer Hipparchus in the 2nd century B.C. Many later astronomers used the data in *Gan Shi Xing Jing* to measure the positions and motions of the sun, the moon and planets.

> 天文历法

中国古代的天文机构除了观测天象之外，还有修订历法、编制历书、历谱并印制发行的职责。早在4300年前，中国古人就已经开始"观象授时"（即观测天象确定时间）。公元前24世纪的尧帝时期就设立了专职天文官，专门从事观象授时的工作。战国时期成书的《尚书·尧典》是中国现存典籍中最早且比较完整地记录观象授时的书籍。当时的天文官确定一年为366日，又以闰月的办法来调配时间，以便使春分、夏至、秋分、冬至四个节气不出差错，从而帮助农民安排农事。

中国古代第一部历法是夏代（约前2070—前1600）禹帝时期的《夏小正》。此历法依据北斗

> Astronomical Calendars

Apart from astronomical observation, the astronomical institutions in ancient China were also responsible for revising calendars, as well as compiling, printing and issuing almanacs and ephemerides. As early as 4,300 years ago, the ancient Chinese already began to determine the time via astronomical observation. During the reign of Emperor Yao in the 24th century B.C., an astronomical official position was specially set to determine the time via astronomical observation. *Shang Shu: Yao Dian*, which was completed in Warring States Period (475 B.C.-221 B.C.), is the earliest book among existing Chinese classics with full record of determining the time via astronomical observation. The astronomers at that time determined that a year had 366 days and that leap month should be used to deploy the time. By

星斗柄所指的方位来确定月份，将一年分为十个月，分别记述了每个月的星象、气象、物象以及所应从事的农事和政事。当时人们对二十八星宿还没有明确的认识，只对"参""辰""昴"几颗天空中肉眼能够观测到的亮星有所记述。

西汉初年采用的历法是秦代颁行的《颛顼历》。但《颛顼历》有一定的误差，不能满足当时农业生产发展的需要。汉武帝下令改定历法，天文学家落下闳等人联合制定新历法，中国第一部较完整的历书

this way, the four solar terms of spring equinox, summer solstice, autumnal equinox and winter solstice would not be led to mistakes, which was helpful for farmers to arrange their farming.

The first calendar of ancient china is *Xia Xiao Zheng*, which was completed during the reign of Emperor Yu in the Xia Dynasty (approx. 2070 B.C.-1600 B.C.). This calendric system divides a year into ten months and defines each month by referring to the position of the Big Dipper's shank. It also describes the star image, weather and natural landscape of each month, as well as the farming and political activities supposed to conduct.

• 北斗星象图（汉 画像石）（图片提供：FOTOE）
Stone Relief of the Big Dipper (Han Dynasty, 206 B.C.-220 A.D.)

● 颁行历书大礼（清）（图片提供：FOTOE）
The ceremony of enacted almanac

《太初历》产生。《太初历》规定一年等于$365\frac{385}{1539}$日，一月等于$29\frac{43}{81}$日。这一规定，从西汉太初元年（前104年）一直用到明末，应用了近两千年。

落下闳在实测的基础上，依照春、夏、秋、冬的顺序，将原来的十月为岁首改为阴历孟春正月为岁首，冬季阴历十二月为岁终，使农事与四季的顺序相吻合，以便于农业生产。《太初历》开始采用有利于农时的二十四节气，并以没有中气的月份为闰月，调整了太阳周天与阴历纪月不相合的矛盾。

People at that time did not have a clear understanding of the 28 Chinese Lunar Mansions, and only recorded a few bright stars that could be seen with naked eye, such as the *Shen* (seven stars included to Orion), *Chen* (Antares) and *Mao* (Pleiades).

The calendar used in the early Western Han Dynasty (206 B.C.-25 A.D.) was the *Zhuanxu Calendar*, which was issued in the Qin Dynasty (221 B.C.-206 B.C.). However, as *Zhuanxu Calendar* had certain errors, it could not meet the development of agricultural production at that time. Then, Emperor Wu of Han Dynasty ordered to revise the calendar. Luoxia Hong and other astronomers formulated a new calendar, China's first relatively complete almanac *Taichu Calendar*. *Taichu Calendar* regulated a year as $365\frac{385}{1539}$ days and a month as $29\frac{43}{81}$ days. This regulation was used for nearly 2,000 years from 104 B.C. to the late Ming Dynasty (1368-1644).

On the basis of actual measurements and in accordance with the order of four seasons, Luoxia Hong changed the beginning of a year from the original 10th month to lunar January (first month of Spring), and changed the end of a year to lunar December (last month

• 老黄历年画

New Year Painting of the Old Calendar

● 老北京人在春节买卖老黄历（图片提供：微图）
Old People in Beijing During the Spring Festival, the Sale of the Old Calendar

《太初历》在天文观测数据的基础上进行推算，形成了一个完整的以地球为中心的宇宙周期系统，又称"落下闳系统"。此宇宙周期系统共分回归年周期、置闰周期、日食周期、干支年周期、干支日周期、木星会合周期、火星会合周期、土星会合周期、金星会合周期和水星会合周期十个基本周期。"落下闳系统"还包括了日月以及五大行星运行的"空间行星背景"，即二十八

of Winter). As a result, farming could coincide with the order of four seasons, which was helpful for the development of agricultural production. It is *Taichu Calendar* that begins the use of the 24 solar terms that is in favor of farming seasons. It sets the month without a second solar term as leap month, adjusting the mismatch between the sun's sidereal revolution and the month system in lunar calendar.

Taichu Calendar makes projections based on the data from astronomical observation and forms a complete earth-centered cosmic periodic system, which is also known as Luoxia Hong's System. The system is divided into ten basic periods: tropical year period, embolism period, solar eclipse period, heavenly-stems and earthly-branches (*Ganzhi*, referring to the lunar calendar) year period, heavenly-stems and earthly-branches day period, Jupiter synodic period, Mars synodic period, Saturn synodic period, Venus synodic period and Mercury synodic period. Luoxia Hong's System also includes the sun and the moon, as well as the five major planets' space background, namely the 28 Chinese Lunar Mansions by which it defines the four seasons. *Taichu Calendar* is China's first calendar with the full text and digital records. It is more accurate and practical than the previous ones.

In the Northern and Southern dynasties (420-589), the famous scientist Zu Chongzhi observed and measured the shadow length of the 23 or 24 days before and after the winter solstice. Then, he averaged the measurements to calculate the date and time of winter solstice. Thus, his method improves the accuracy of the calculation. He measured the length of the tropical year with a tablet meter

• 落下闳像
Portrait of Luoxia Hong

宿，根据天空中二十八宿的位置可以判定春、夏、秋、冬四季。《太初历》是中国第一部有完整文字和数字记载的历法，较之前的历法更加准确、实用。

南北朝时期，著名科学家祖冲之观测冬至前后二十三四天日影长度，再取平均值，求出冬至发生的日期和时刻,这一方法提高了测定冬至时刻的精度。他用圭表测定回归年的长度,用浑仪等测角器测定太阳在恒星间的位置,研究太阳在一年中运动的快慢变化规律,测定出冬至点逐年变化的数值。公元510年,祖冲之生前制定的当时最科学的历法《大明历》颁行。《大明历》确定一年为365.24281481天,与现代天文所测结果仅有50余秒的误差。

(*Guibiao*, an ancient Chinese sundial), and determined the position of the sun among fixed stars with armillary sphere and other goniometer. Also, he studied the law of the speed change of the sun's motion in a year, and measured the year-by-year changing values of the winter solstice point. In 510, *Daming Calendar*, which was the most scientific calendar at that time and was formulated during Zu Chongzhi's lifetime, was enacted. It defines a year with 365.24281481 days, which deviates only 1/600000 to modern astronomical measurement.

天文仪器的制作

早在公元前7世纪,中国就已经出现了通过测量日影长度以定方向、节气和时刻的天文仪器——圭表。其原理为:长直的标杆(即圭)竖直立起,放于地面刻板(即表)的正南面,根据太阳照射时投影在地面刻板上的长度来判断时间。正因为有了圭表,古人才很早就知道了一年为365天。除了圭表,还有在其基础上发展而来的日晷。

汉代天文学的集大成者张衡著有《灵宪》一书,书中系统地描述了浑天说的宇宙模型。同时,他在前人研究基础上制造了水运浑象。浑象是用来演示天体运动的。

张衡制造的水运浑象对后世浑象的制造影响极大。公元1085年,北宋天文学家苏颂等人开始设计制作水运仪象台,前后历时3年。仪象台以水力运转,集天象观察、演示和报时三种功能于一体,是一座上狭下广、正方形的木结构建筑，大约有12米高，7米宽，共分为3大层。上层是一个露天的平台，设有浑仪一座，用龙柱支持，下面有水槽。浑仪上面覆盖有防止日晒雨淋的木板屋顶，为了便于观测，屋顶可以随意开闭。中层是一间没有窗户的密室，里面放置着浑象。下层包括报时装置和全台的动力机构等，设有向南打开的大门，门里装置有5层木阁，木阁后面是机械传动系统。水运仪象台采用了民间使用的水车、筒车、凸轮和天平秤杆等机械工具，把观测、演示和报时设备集中起来，组成了一个整体，成为一个自动化的天文台。英国的李约瑟博士等人认为，水运仪象台是"欧洲中世纪天文钟的直接祖先"，并称赞苏颂是中国古代甚至是中世纪世界范围内最伟大的博物学家和科学家之一。

Production of Astronomical Instruments

As early as the 7th century B.C., the tablet meter (*Guibiao*, an ancient Chinese sundial) has already appeared, which is an astronomical instrument used for indicating the time,

• 日晷
Sundial

• 圭表
Tablet Meter

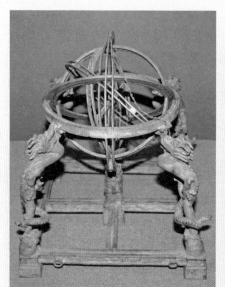

• 浑仪
Armillary Sphere

• 星图节气天文钟（清）
Astronomical Clock with Star Charts and Solar Terms (Qing Dynasty, 1616-1911)

direction, and season by measuring the length of shadow. Its working principle is that the long and straight pole (the tablet) stands vertically to the south of the ground dial (the meter), so that the time can be determined by the shadow of the pole projected on the ground dial. Thanks to tablet meter, the ancients were able to find very early that there are 365 days in a year. Besides tablet meter, another similar instrument is sundial, which evolved from tablet meter.

Zhang Heng, an astronomy master in the Han Dynasty (206 B.C.-220 A.D.), systematically described the cosmological model of the Theory of Sphere Heaven in his book *Ling Xian*. Meanwhile, based on the astronomical studies of the pioneers, he created a water-driven celestial globe. The celestial globe is used to demonstrate the movements of celestial bodies.

Zhang Heng's water-driven celestial globe has exerted great influence on the making of the later celestial globes. In 1085, Su Song and other astronomers of the Northern Song Dynasty (960-1127) began to design and build the water-driven astronomical clock tower for 3 years. Driven by water, the astronomical clock tower has the functions of astronomical observation and demonstration as well as timekeeping. It is a wooden tower in square shape with the body narrowing from the bottom to top. It has three major stories with a height of about 12 meters and a base width of about 7 meters. The top story is an open-air terrace where an armillary sphere is placed. The armillary sphere is supporting by dragon pillars, and the water channels beneath it are used to determine the horizontal direction. The armillary sphere is covered with a wooden roof that used to keep out the sun and rain, and in order to facilitate the observation, the roof can be opened and closed. The middle story is an adytum without windows, where a celestial globe is placed. The ground story contains the timekeeping devices as well as the actuating units of the entire tower. Inside of this story's southward door, there is a 5-layer wooden cabinet, behind which is the mechanical transmission system. By adopting the mechanical tools of civilian waterwheel, watermill, cam, balance arm and so on, the water-driven astronomical clock tower unites

● 苏颂制水运仪象台（模型）

（图片提供：微图）

Su Song's Water-driven Astronomical Clock Tower (Model)

the devices for observation, demonstration and timekeeping as a whole, which becomes an automated observatory. Dr. Joseph Needham from British and other scholars believed that "the water-driven astronomical clock tower is the direct ancestor of the astronomical clocks of medieval Europe". They also praised Su Song as the greatest naturalist and scientist in ancient China and even over world during the middle ages.

地理学
Geography

"地理"一词最早出现在《易经》当中,有"仰以观于天文,俯以察于地理"的记载。"上知天文、下知地理"也逐渐成为古人对博学多识的人的一种描述。中国古代地理学起步很早,在地图制作、学术著作等方面有突出成就,在先秦时期中国人就已经撰写了地理学著作《山海经》。

Dili, the word meaning geography in Chinese, first appeared in *The Book of Changes*, which had a record of "looking up to observe astronomy and looking down to survey geography (*Dili*)". Knowing both astronomy and geography then gradually became ancient people's description to knowledgeable persons. Ancient Chinese geography started very early and had outstanding achievements in map-making, academic works and other aspects. Before the Qin Dynasty (221 B.C.–206 B.C.), Chinese people had written a geography book, *The Book of Mountains and Seas*.

> 古代地理学概述

地理学是伴随着人类的生产活动而产生的，随着社会的发展，人们在不同时期对地理学有着不同程度的认识。

> Overview of Ancient Geography

Geography was born along with human activities. With the development of society, people at different times had different levels of knowledge on geography.

In primitive society, people began to conceive the concept of geography through activities of fishing, hunting, collecting and constructing. They began to accumulate the knowledge of terrains, rocks, rivers, and other geographic

• 新石器时期陶器
陶器是以黏土为主要原料烧制而成的，这表明当时人们已经对土壤的性能有了一定的认识。

Pottery of the Neolithic Age (approx. 8,000 years ago)

Pottery is fired with the main raw material of clay, which shows that people at that time have had a certain understanding of the property of soil.

● 甲骨文 "风" "雨"
Oracle Characters Meaning Wind and Rain

在原始社会时期,人们在渔猎、采集、建筑等活动中萌生了地理概念,初步积累了对于居住地附近的地形、岩石、河流等地理要素的认识。进入氏族社会时期,人们已经形成了地理方位概念,在距今五千年的红山文化遗址中,考古工作者发现了按照南北轴线对称分布的祭坛,这反映了当时的天地观和地理学思想。

殷商时期,随着最原始的文字——甲骨文的出现,古人对地理的发现得以用文字的方式记录下来。在这一时期,已经出现了"南""北"等文字。周代,人们对于地理更加重视,出现了司徒、

features nearby their residence. And by the period of clan society, the concept of geographic orientation was formed. In the Hongshan culture sites dating back to 5,000 years ago, archaeologist found an altar that was symmetrically distributed in accordance with the north-south axis, which reflects the concept of heaven and earth as well as the geographical thoughts at that time.

In the Shang Dynasty (1600 B.C.-1046 B.C.), with the advent of the most primitive writing Oracle, the ancients were able to record geographical discoveries by writing. During this period, characters about south, north and other words appeared. In the Zhou Dynasty (1046 B.C.-256 B.C.), more attention

司马等官职，其中司徒主要从事土地和农业生产的管理工作。

战国时期，魏国人假借大禹之名撰写了《尚书·禹贡》，其中首次提出"九州"这个地理概念，即以山脉、河流为标志，将国家分为九个行政区。书中对九个行政区的疆域、山脉、河流、植被、土壤、少数民族、交通等内容进行了简要描述。齐国人邹衍在九州概念的基础上，提出了"大小九州"的地理学概念。他认为中国之内有九州，为小九州，中国之外也有九州，为大九州，而中国只占天下的八十一分之一。对此，《史记·孟子荀卿列传》中有相关记载："中国名曰赤县神州，赤县神州内自有九州……中国外如赤县神州者九，乃所谓九州也……中国者，于天下乃八十一分居其一分耳。"

经过了长久的积累，地理学在秦汉时期作为一种自然科学学科初步形成。西晋时，裴秀主持绘制了中国迄今为止最早的地图集——《禹贡地域图》。在此书中，裴秀提出了绘制地图的六项原则——"制图六体"，并强

was attached to geography by setting up official positions, such as Situ, an official in charge of the management of land and agricultural production.

During the Warring States Period (475 B.C.-221 B.C.), people of the State of Wei wrote the article *Shang Shu: Yu Gong* in the name of Yu the Great (an emperor of ancient China). This article put forward the geographical concept of Nine States (*Jiuzhou*, a poetic name for China) for the first time in the history of China. The Nine States referred to ancient China's nine administrative regions marked by mountains and rivers. The article briefly described the boundaries, mountains, rivers, vegetation, soil, ethnic groups, traffic and other features of the nine administrative regions. Based on this concept of Nine States, Zou Yan from the State of Qi put forward the geographic concept of the Big and Small Nine States. He believed that there were nine small states inside China while there were nine big states outside China, and that China took one eighty-first of the world. In this regard, relevant records can be found in *Historical Records: Biographies of Mencius and Xuncius*: "China is called the Divine Land of Red County, inside

- 马王堆帛书地图（汉）（图片提供：FOTOE）
  此地图对当时的主要城邑、山脉位置及河流的走向都有十分准确的描述。
  Silk Manuscript Map Unearthed from Mawangdui (Han Dynasty, 206 B.C.-220 A.D.)
  This map has a very accurate description of the main cities, the location of mountains and the direction of rivers at that time.

调制图六体必须要综合运用，他被称为"中国传统地图学的奠基人"。制图六体是指分率（即比例尺）、准望（即方向）、道里（即距离）、高下（即取下为水平直线距离）、方邪（即取斜为直线距离）和迂直（即取直为直线距离）。汉代时，史学家班固首次将阐述疆域地理知识的地理志编入正史之中。此外，在长沙马王堆出土的汉代帛书地图是中国现存的最早地图，该地图采用了固定的比例尺，代表了当时的最高水平。

唐代是中国历史上经济、文化等方面大繁荣的鼎盛时期之一，随着国内外交流愈加紧密，人们对地理的认识也从本国扩展到周边国家。从801年开始，在宰相、地理学家贾耽的主持下，历时17年，中国古代最早的世界地图——《海内华夷图》问世。贾耽在裴秀制图六体的基础上，采用古今对照、双色绘画的方式绘制了该地图。同时，佛教在中国逐渐推广，也为人们认识周边国家作出了重要贡献，其中最具代表性的事件为玄奘西行印度、

which there are nine states… Outside China, there are nine such Divine Lands of Red County, naming the Nine States… China is one in eighty-one of the world."

After a long-term accumulation, geography was initially developed to be a natural science discipline in the Qin and Han dynasties. In the Western Jin Dynasty (265-317), Pei Xiu drew and edited so far China's earliest atlas, *Yu Gong's Geographical Maps*. In this book, Pei Xiu brought forward six principles for map drawing, namely the Six Elements of Map, which he believed should be made comprehensive use. So, he was known as the "founder of Chinese traditional cartography". The six elements referred to the scale, direction and distance, as well as taking bottom line as the horizontal distance of heights, taking oblique line as the horizontal distance of rectangles, and taking straight line as the horizontal distance of detours. In the Han Dynasty (206 B.C.-220 A.D.), historian Ban Gu for the first time compiled geographical chronicles about the geographical knowledge of territory into official history. Moreover, the silk manuscript map of the Han Dynasty (206

- 《郑和航海图》之《针路图》

（图片提供：FOTOE）

出自明代茅元仪《武备志》。依靠指南针测得的航线名为"针路"，记载针路的书籍为"针经""针谱"。

Compass Map in *Zheng He's Nautical Chart*

This map comes from *Treatise on Armament Technology* written by Mao Yuanyi in the Ming Dynasty (1368-1644). The ship routes measured by compass were called Compass Routes in ancient China, and the books recording compass routes were called Book of Compass or Chart of Compass.

B.C.-220 A.D.) unearthed from Mawangdui in Changsha is China's earliest existing map. Adopting a fixed scale, the map represents the highest level of drawing at the time.

The Tang Dynasty (618-907) is one of the most prosperous heydays for economy, culture and other aspects in Chinese history. With closer domestic and international exchanges, people's knowledge of geography at the time expanded from China to neighboring countries. From the year of 801 and under the auspices of the prime minister and geographer Jia Dan, it took 17 years to finish ancient China's earliest world map, *Map of China and Other Countries*. On the basis of Pei Xiu's Six Elements of Map, Jia Dan drew this map by means of ancient-modern comparing and two-color painting. Meanwhile, the gradual spread of Buddhism in China also made important contributions to people's understanding of neighboring countries. The most representative cases include Xuanzang's westward journey to India

• 郑和宝船模型（图片提供：微图）
Model of Zheng He's Treasure Ship

鉴真东渡日本。

　　明代，郑和带领船队下西洋，不仅是中国航海史上的壮举，更是一次伟大的地理大发现。从明代永乐三年（1405年）至宣德八年（1433年），郑和奉命率领船队

and Jianzhen's eastward voyage to Japan.

　　In the Ming Dynasty (1368-1644), Zheng He and his fleet's voyage to the western seas was not only a feat of China's maritime history, but also a great geographical discovery. From 1405 to 1433, Zheng He received orders to

七下西洋。据《明史·郑和传》记载，郑和下西洋共经过30多个国家或地区，包括东南亚的爪哇、泰国、苏门答腊、印度半岛南端以及非洲东岸的红海、麦加等地。郑和航海的探索比西方著名海上探险家达·伽马、哥伦布等人早了80多年，代表了中国古代海上探索的最高成就。

lead fleets to western seas for 7 times. According to *History of the Ming Dynasty: Biography of Zheng He*, Zheng He's voyages went by more than 30 countries or regions, including Java, Thailand, Sumatra and the southern tip of the Indian Peninsula in Southeast Asia, as well as Red Sea, Mecca and other places in the east coast of Africa. Zheng He's maritime exploration, which represents the highest achievements of maritime exploration in ancient China, is 80 years earlier than that of Vasco da Gama, Christopher Columbus and other famous western maritime explorers.

> 古代地理学著作

在先秦著作《山海经》一书中，古人开始按照由南向西，再依次向北、向东的方位有序地记载地理现象。此后，各个时期都有突出的地理学成就，涌现了许多著名的地理学家及丰富的地理学著作。

郦道元与《水经注》

郦道元是北魏时期（386—534）著名的地理学家。少年时代，他随父亲到过山东，还经常与朋友一起到有山有水的地方游览，观察水流情况。后来，郦道元在山西、河南、河北等地区做官，经常趁闲暇之时进行实地考察和调查，日积月累，掌握了许多有关地理的原始资料。

除了进行实地考察，郦道元还

> Ancient Works on Geography

In *The Book of Mountains and Seas*, the ancients before the Qin Dynasty (221 B.C.-206 B.C.) began to record geographical phenomenon orderly in accordance with the direction from south to west, then to north and east in sequence. Since then, prominent achievements were made in the field of geography during various periods, and many famous geographers and rich geographical works sprung up.

Li Daoyuan and *Commentary on the Waterways Classic*

Li Daoyuan was a famous geographer of the Northern Wei Dynasty (386-534). In his boyhood, he had been to Shandong Province with his father. In addition, he often visited mountains and waters with friends to observe natural

● 《水经注》（清刻本）
Commentary on the Waterways Classic
(Block-printed Edition, Qing Dynasty, 1616-1911)

十分专注于阅读地理方面的书籍。在阅读《水经》（《水经》是中国第一部记述河道水系的地理专著，全书约一万字，共记述了137条主要河流的水道情况）时，他发现虽然该书对大小河流有相对准确的记载，但是由于时代的更替，许多河流改道，河流的名称也有所变化。于是，郦道元根据自己亲身考察所得到的资料以及大量历史文献资料，为《水经》作注，补充了许多河流的实时信息及相关内容。郦道

sceneries. Later, when Li took the office in Shanxi, Henan, Hebei and other provinces, he often spent his spare time on field surveys and investigations. Day by day, he collected many raw data on geography.

In addition to the field surveys, Li was also very enthusiastic about reading geographical books. When reading *Waterways Classic* (China's first geographical monograph recording river systems, which has a total of about 10,000 Chinese characters and a total records of 137 major rivers' waterways), he found that although the book had relatively accurate records on the major courses of rivers, many of them had made diversion or many of their names had been changed due to the replacement of the times. Therefore, according to the data from personal surveys and based on the reference to large number of historical literatures, he added notes to the *Waterways Classic* by supplementing the real-time information and relevant content of the rivers. Li's *Commentary on the Waterways Classic* had a total of 40 volumes and over 300,000 Chinese characters, describing hydrological conditions of 1,252 rivers. Meanwhile, he also described in detail the geology,

● 三峡（图片提供：全景正片）

三峡是长江干流中的一段大峡谷，为中国十大风景名胜区之一。三峡西起重庆奉节的白帝城，东至湖北宜昌的南津关，由瞿塘峡、巫峡、西陵峡组成，全长约193千米。

The Three Gorges

The Three Gorges is a huge canyon along the Yangtze River and is one of China's top ten scenic spots. The Three Gorges stretches from Baidi Town in Fengjie County of Chongqing in west to Nanjin Pass in Yichang City of Hubei Province in east, with a total length of 193 km. It consists of Qutang Gorge, Wu Gorge and Xiling Gorge.

元的《水经注》共40卷，30多万字，记述了1252条河流的水文情况，同时，他还对河流两岸的地质、地貌、气候、民俗、古迹及神话传说等做了详细的描述。

《水经注》不仅是一部丰富的地理学著作，还是一部出色的游记著作。在描写三峡时，郦道元写道："自三峡七百里中，两岸连山，略无阙处。重岩叠嶂，隐天蔽日，自非亭午夜分，不见曦月。至于夏水襄陵，沿溯阻绝。或王命急宣，有时朝发白帝，暮到江陵，其间千二百里，虽乘奔御风，不以疾也。春冬之时，则素湍绿潭，回清倒影。绝巘多生怪柏，悬泉瀑布，飞漱其间，清荣峻茂，良多趣味。"这段文字既写出了三峡山势高、重山相连且中间狭窄的特点，说明了夏季江水上涨、江上船只受阻的情景，还将三峡春冬时节江水清澈、树木繁荣、草木茂盛的景色描绘得淋漓尽致，此段有关三峡的描写堪称全书最优美的文字。

此外，有关北京城的最早资料就是出自郦道元的《水经注》。书中记载了当时北京城的城址、近郊

topography, climate, customs, monuments as well as the myths and legends on both sides of the rivers.

The *Commentary on the Waterways Classic* is not only a rich geography book, but also a remarkable travel literature. In the description of the Three Gorges, Li wrote that: "Along the 350 km river of the Three Gorges, rolling hills on both sides line up without any interruption. Layers and layers of rocks and cliffs are blotting out the lights. You cannot see the sun or noon until midday or midnight. In summer, the river diffuses on the hills on both sides, blocking the traveling boats. Sometimes when the emperor has urgent edicts to convey, messenger can reach Jiangling County at dusk by boat if he starts from Baidi Town at morning. The two places are 600 km apart, so traveling by boat is faster than by horse. In spring and winter, white riffles flow with clear whirling ripples, and green deep pools reflect various sceneries. Grotesque cypresses insert into sheer peaks, waterfalls in all sizes rush out from steep cliffs, and clear water is surrounding mountains with exuberant trees and grasses. What a tasted picture!" This text not only traces out the Three Gorges' features of narrowly connected high hills,

的历史遗迹、河流和湖泊的分布状况以及北京地区的人们早期所进行的一些大规模的改变自然环境的活动，例如拦河堰的修筑、天然河流的导引和人工渠道的开凿等。

沈括与《梦溪笔谈》

沈括（1031—1095）是北宋时期著名的科学家。他出生于官宦世家，从小便跟随父亲四处游历，饱览大山大河以及各地的风土人情。他一生四处为官，几乎走遍了大半个中国，晚年时退出政坛，隐居在江苏镇江的梦溪园。他不但对指南针有研究，而且还最早发现了地磁偏角的存在。沈括依据自己一生的研究，撰写了一部笔记体著作，即《梦溪笔谈》。

《梦溪笔谈》的内容十分丰富，涉及天文、历法、气象、地质、地理、生物、农业、水利、建筑、医药、历史、文学等诸多领域，"石油"这一名称，就出自《梦溪笔谈》，并一直沿用至今。《梦溪笔谈》一书集中反映了中国当时科技发展的最新成就，称得上是中国科学史上的里程碑。《梦溪

but also describes the scenario that the river water rises to block traveling boats in summer. What's more, the clear water, flourish trees and luxuriant grasses of the Three Gorges in spring and winter are depicted incisively and vividly. This text about the Three Gorges is said to be the most beautiful writing in the book.

In addition, the earliest data about Beijing is from Li's *Commentary on the Waterways Classic*. It has records about the position of Beijing City, the historic sites in its outskirts, and the distribution of rivers and lakes. Some large-scale nature-transforming activities conducted by early people in Beijing area are also recorded, such as the construction of the weirs, the diversion of natural rivers, and the excavation of artificial channels.

Shen Kuo and *Dream Pool Essays*

Shen Kuo (1031-1095) was a famous scientist of the Northern Song Dynasty (960-1127). Thanks to his aristocratic family, Shen was able to visit lots of places with his father at the childhood, enjoying the mountains and rivers as well as the customs of various places. When he was an official, he wandered around almost half of China. When he retired in his later years, he lived in seclusion

- **沈括像**

  宋代杰出的科学家，于天文、音乐、医药、卜算均有所建树，著书甚多，《梦溪笔谈》是他的代表作。

  Portrait of Shen Kuo

  He is an outstanding scientist of the Song Dynasty (960-1279), who accomplished much in astronomy, music, medicine and divination. He has written a lot of books, and *Dream Pool Essays* is his masterpiece.

笔谈》被西方学者赞誉为"中国古代的百科全书"，英国科学史家李约瑟赞誉《梦溪笔谈》为"中国科学史上的坐标"。

沈括根据自己对浙江雁荡山、陕北黄土高原地貌地质的实地考察，明确提出流水有侵蚀作用的说法。在地图制作方面，他以熔蜡和木屑制作出世界第一个立体地图，这一发明比欧洲早约700年。

in the Dream Pool (*Mengxi*) Park of Zhenjiang City in Jiangsu Province, and concentrated himself on studies. In addition to the study of compass, he was the first person to discover the existence of magnetic declination. Based on his life-time study, he wrote a sketchbook—*Dream Pool Essays*.

Dream Pool Essays has very rich contents in astronomy, calendar, meteorology, geology, geography, biology, agriculture, water conservancy, architecture, medicine, history, literature, and many other fields. It is *Dream Pool Essays* that the Chinese word *Shiyou* (petroleum) comes from, which has been in use ever since. This book epitomizes the highest achievements of China's scientific and technological development at that time. Thus, it is regarded as a

徐霞客与《徐霞客游记》

徐霞客（1587—1641），明末地理学家、旅行家。徐霞客的父亲一生不愿为官，也不愿同权贵交往，喜欢到处游览，欣赏山水景观。受父亲的影响，徐霞客自幼就喜欢读历史、地理和探险、游记之类的书籍。

《徐霞客游记》是徐霞客30

milestone in the history of Chinese science. In the eyes of many Western scholars, it is praised as the "encyclopedia of ancient China". And according to the praising words of British historian Joseph Needham, it is the "coordinate of the history of Chinese science".

According to his field surveys on the landform and geology of Yandang Mountain in Zhejiang Province as well as the Loess Plateau in northern Shaanxi

● 浙江雁荡山 （图片提供：微图）
雁荡山由火山喷发而成，有"古火山立体模型"之称，2002年被联合国教科文组织评为世界地质公园。
Yandang Mountain in Zhejiang Province
Yandang Mountain was formed by volcanic eruption. Regarded as the "three-dimensional model of ancient volcano", it was ranked as a World Geopark by UNESCO in 2002.

多年旅行考察的真实记录，是他一生最杰出的作品，开创了地理学上系统地观察自然、描述自然的新方向。该书既是考察中国地貌地质的地理名著、描绘华夏风景的旅游巨著，也是文字优美的文学佳作，在中国文化史上具有深远的影响，被人们誉为"古今游记第一"。

徐霞客是世界上最早对石灰岩进行系统考察的地理学家。在广西、贵州、云南等地区，他亲自深入考察的洞穴就有270多个，并且对其中一些洞穴中的喀斯特地貌进行了解释。其中，对广西桂林七星岩的记叙尤为详尽，他不仅对岩洞进行了多方位的考察、多角度的

Province, Shen clearly put forward the diction of Runoff Erosion. In the field of cartography, he produced the world's first three-dimensional map with melting wax and sawdust. This invention is about 700 years earlier than that in Europe.

Xu Xiake and *Travels of Xu Xiake*

Xu Xiake (1587-1641) was a geographer and traveler of the late Ming Dynasty (1368-1644). Xu's father in his life time was reluctant to be an official or to work with bigwigs, but liked to admire landscapes and views by traveling. Influenced by his father, Xu had been fond of reading books about history, geography, adventure and travels since he was a child.

Travels of Xu Xiake is a real record of Xu's travels and surveys of over 30 years. This book is the most outstanding work in his life, which has created a new direction of systematically observing and describing the natural in geography. It is a remarkable geographical book investigating China's landform and

• 徐霞客像
Portrait of Xu Xiake

geology, an outstanding tourist work depicting China's landscapes, and an excellent literary writing utilizing beautiful words. With a profound effect in the history of Chinese culture, it is praised as the No.1 Travels at All Times.

Xu was the world's first geographer to have a systematic investigation on limestone. In Guangxi, Guizhou, Yunnan and other areas, he inspected personally more than 270 caves, and explained the phenomenon and land feature in some of the caves, namely karst. His description of the Seven Star Cave in Guilin City of Guangxi Zhuang Autonomous Region was in considerable detail. In addition to the multi-direction investigation and multi-angle description on the cave, he also had an accurate judgment of its topographic feature, cave structure and forming reasons. He pointed out that the phenomenon of karst was caused by the dissolution happened during running water's washing against limestone. This is the most important geographical discovery in *Travels of Xu Xiake*.

● 七星岩（图片提供：图虫创意）
Seven Stars Cave

描绘，而且对地形特征、岩洞结构及形成原因都有准确的判断。他指出，正是由于流水对石灰岩进行冲刷的过程中发生的溶蚀作用，才产生了岩溶现象，这是《徐霞客游记》中最重要的地理学发现。

四大发明

"四大发明"这一说法最早由英国科学家、历史学家李约瑟提出,指的是中国的造纸术、印刷术、指南针和火药。英国哲学家培根也曾说过:"印刷术、火药、指南针这三种发明已经在世界范围内把事物的全部面貌和情况改变了。""四大发明"的出现,不仅推动了中国历史、文化、经济等方面的发展,甚至推动了全世界的进步。

造纸术。中国是世界上最早养蚕的国家,古人以蚕茧抽丝织绸,漂絮后,席上会留下一些残絮,晒干之后,被称为"方絮"。这种纸张十分粗糙,不能用来写字,因此并没有在文化传播上发挥积极的作用。东汉(25—220)时,蔡伦对造纸术进行了改进,选用树皮、麻头、破布、旧渔网等方便获得的原料,经过浸、捣等工序制成纸浆,再经过抄、烘等步骤,将纸浆晾干之后,就制造出更适合书写的纸张。造纸术作为中国最伟大的发明之一,充当了人类文明、科技传播的"使者"。造纸术首先传入与中国毗邻的朝鲜和越南,随后传到了日本,而后又传入阿拉伯,再经阿拉伯地区传入欧洲。到19世纪,中国的造纸术已传遍世界各国。

● 蚕茧
Silkworm Cocoon

Four Great Inventions

The diction of the Four Great Inventions was first put forward by British scientist and historian Joseph Needham, referring to China's papermaking, printing, compass and gunpowder. British philosopher Francis Bacon once said: "The three inventions of printing, gunpowder and compass have changed the outlook and situation of things all over the world." The emergence of the Four Great Inventions has promoted not only China's developments in history, culture, economy and other aspects, but also even the progress of the whole world.

Papermaking. China is the world's first nation to breed silkworms for silk weaving. The ancient Chinese reeled off raw silk from silkworm cocoons for silk weaving. After being rinsed, some residuary wadding was left on mat. By drying in the sun and then stripping down from the mat, such paper was called Rectangle Wadding (*Fangxu*). Since such paper was too rough to be used for clear writing, it did not play an active role in cultural transmission. In the Eastern Han Dynasty (25-220), papermaking was improved by Cai Lun. He selected and used barks, hemp wastes, rags, old fishnets and other easily obtained raw materials, which would be treated in steps of dipping, smashing, paper-shaping, and baking. After the pulp was air-dried, the paper more suitable for writing eventually was made. As one of China's greatest inventions, papermaking acted as a "messenger" for the transmission of human civilization and technology. Papermaking was first introduced into the adjacent North Korea and Vietnam, and then spread to Japan and Arabia, and was brought to Europe via the Arab region. By the 19th century, China's papermaking has spread to countries all over the world.

• 《蔡伦造纸图》
Picture of Cai Lun Making Paper

- **雕版印刷**
  Woodblock Printing

　　印刷术。早在隋唐时期，雕版印刷就大行于市。雕版印刷，即在一定厚度的平滑木板上，粘贴上抄写工整的书稿，将稿纸正面和木板相贴，雕刻工人把版面没有字迹的部分削去，这样雕版就做成了。印刷时，在凸起的字体上涂上墨汁，然后把纸覆在上面，以手拂拭纸背，字迹就留在纸上了。但是，雕版上出现错字很难修改，错字的出现常常使整块雕版都要被废弃。北宋庆历年间（1041—1048），印刷工人毕昇发明的活字印刷术改变了这一情况。他总结历代雕版印刷的经验，经过反复实验，制成了泥活字，完成了印刷史上的一次伟大革命。毕昇的活字印刷术比德国谷登堡活字印刷术早了约400年。

　　Printing. As early as the Sui and Tang dynasties , woodblock printing was used on a large scale in China. Woodblock printing was to past neatly-written manuscript on a smooth woodblock with a certain thickness. With the front side of the manuscript paper pasted on the woodblock, the parts on the woodblock without characters would be cut off to form raised characters. In order to print, the raised characters would be coated with ink and was covered with paper. By whisking off the back side of the paper, the characters were finally printed on the paper. However, as it was difficult to correct the wrong characters on the woodblock, the entire woodblock was often abandoned due to wrong characters. During the Reign of Qingli (1041-1048) of the Northern Song Dynasty (960-1127), typographer Bi Sheng invented the movable type printing to change this situation. By summing up the experience of the previous woodblock printings, Bi created the clay type after various experiments, which was a great revolution in the history of printing. Bi's movable type printing appeared about 400 years earlier than that of Johannes Gutenberg in German.

火药。古代统治者将长生的希望寄托在炼丹家身上，希望他们能够炼制出长生不老药，这一需求使得炼丹风气盛行。在炼制丹药的过程中，一次偶然的爆炸使得火药这一重要物质被发明。唐代（618—907）时，炼丹家已经掌握了一种重要的经验，即硫、硝、炭三种物质混合在一起遇到明火就会发生爆炸。这一特性引起了古代军事家的兴趣，并将其应用到战争中，制造了许多火药武器，例如火箭。火药的使用改变了古代战争的作战方式，结束了冷兵器时代。

Gunpowder. The ancient Chinese rulers placed their hope of longevity on alchemists, hoping them could refine the elixir of life. This demand pushed forward the prevalence of alchemy at the time. In the process of elixir refining, an accidental explosion brought the discovery of the important invention of gunpowder. In the Tang Dynasty (618-907), alchemists mastered an important experience that the mixture of sulfur, saltpeter and charcoal would explode if it was ignited with open flames. This feature aroused the interest of ancient militarists, who applied it to wars by creating a lot of gunpowder weapons, such as rockets.

• 威远将军炮（清）
General Weiyuan's Cannon
(Qing Dynasty, 1616-1911)

The use of gunpowder changed the ways of fighting in ancient wars, and made an end of the era of cold weapons.

指南针是人们判别方向的重要工具，世界上最早的指南针——司南在战国时期（前475—前221）便已出现。"司南"是用天然磁石制成的，样子像一把汤勺，可以放在平滑的底盘（方形的金属盘）上，并且可以自由旋转。当它静止的时候，勺柄就会指向南方。除了司南，北宋时期（960—1127）还出现了一种更为便捷的指南工具——指南鱼。指南鱼由薄铁片制成，一般两寸长、五分宽，鱼的腹部略下凹，像一只小船，磁化后浮在水面，就能指示南北。由于液体的摩擦力比固体小，转动起来比较灵活，指南鱼比司南更灵敏，更准确。在北宋时期的一部军事著作《武经总要》中，就有关于指南鱼在行军中辨明方向的记述。指南针在军事、日常生活、地形测量以及航海上都发挥了极大作用。12世纪以后，指南针传到了阿拉伯国家和欧洲，大大推动了世界航海事业的发展。

Compass. As an important tool for people to discriminate directions, the world's first compass, which was called South-pointing Ladle (*Sinan*), appeared in China in the Warring States Period (475 B.C.-221 B.C.). Made of lodestone, the South-pointing Ladle looked like a ladle, which could maintain a balance on a smooth chassis (a square metal plate). It could also rotate freely, and when it stopped, the handle of the ladle would point to the south. In addition to South-pointing Ladle, a more convenient guiding tool South-pointing Fish appeared in the

• 司南
South-pointing Ladle

Northern Song Dynasty (960-1127). The South-pointing Fish was made of thin iron sheet cut into a fish shape. With a length of about 6.667 cm and a width of about 1.667 cm, the fish had a slightly concave belly, which looked like a boat. After the magnetization treatment, it would point to the south and north when floating on water surface. As the friction of liquid was smaller than solid, the South-pointing Fish could make a more flexible turning, and was more sensitive and accurate than the South-pointing Ladle. *Military Pandect*, a military book of the Northern Song Dynasty (960-1127), had records about the use of South-pointing Fish for direction discrimination in marches. Compass played a significant role in military, daily life, topographic survey, and especially navigation. After the 12th century, compass spread to the Arab countries and Europe, which greatly promoted the development of the world's maritime industry.

• 指南车（图片提供：微图）
相传指南车由黄帝发明，在黄帝与另一个部落首领蚩尤作战时发挥了重要作用。
South-pointing Cart
According to legends, the south-pointing cart was invented by the Emperor Huang. It played an important role in Emperor Huang's fight with another tribal leader Chiyou.

数学
Mathematics

在中国古代,数学叫做"算术",又称"算学"。数学是儒家学者必须掌握的六艺(六艺即六项基本才能,分别为礼、乐、射、御、书、数)之一。春秋时期,人们就已经开始应用十进位值制记数法,并谙熟九九乘法表、整数四则运算,还开始使用分数。

In ancient China, mathematics was called arithmetic, which was included as one of the Six Arts (six basic talents, i.e. rites, music, archery, charioting, calligraphy and mathematics) that a confucianist should mastered. In the Spring and Autumn Period (770 B.C.–476 B.C.), Chinese people started to apply the decimal system and the fraction and mastered the multiplication table and integer arithmetic.

> 算筹与筹算

春秋时期，中国古人普遍使用竹木制的算筹进行加、减、乘、除、开方等运算，这种方法称为"筹算"。算筹是一根根同样长短、同样粗细的小木棍，一组有270根左右，人们平时随身携带，在需要计算的时候拿出来使用。算筹是算盘出现之前最主要的算数工具。在使用算筹进行计算时，小木棍的摆放是有规律的，表示个位数时，1至5竖着摆，数字是几就摆几根小木棍；6、7、8、9则分别在1、2、3、4的小木棍上面横着放置小棍子；表示多位数时，个位竖着摆，十位横着摆，百位竖着摆，以此类推，遇零就置空。

中国古代十进位值制的算筹记数法在世界数学史上是一个伟大的

> Counting Rods and Rod Arithmetic

In the Spring and Autumn Period (770 B.C.-476 B.C.), ancient Chinese usually used counting rods made from bamboo or wood to do the calculations of addition, subtraction, multiplication, division and radication, which was called "rod arithmetic". These counting rods were wood sticks with same length and thickness and constituted a set with about 270 sticks. It was easy to carry around and convenient to operate. The counting rods were the most significant calculation tool before the invention of abacus. There were several rules regarding to the operation of rods: as to the units digit, the numbers from one to five were presented by the rods with corresponding number; the numbers from six to nine needed to add horizontally placed sticks above the numbers from one to four. As to the multidigit number, the units digit was

1×1=1								
1×2=2	2×2=4							
1×3=3	2×3=6	3×3=9						
1×4=4	2×4=8	3×4=12	4×4=16					
1×5=5	2×5=10	3×5=15	4×5=20	5×5=25				
1×6=6	2×6=12	3×6=18	4×6=24	5×6=30	6×6=36			
1×7=7	2×7=14	3×7=21	4×7=28	5×7=35	6×7=42	7×7=49		
1×8=8	2×8=16	3×8=24	4×8=32	5×8=40	6×8=48	7×8=56	8×8=64	
1×9=9	2×9=18	3×9=27	4×9=36	5×9=45	6×9=54	7×9=63	8×9=72	9×9=81

● 九九乘法表
Multiplication Table

创造。比起其他古老文明，中国的算筹记数法的优势十分明显。古罗马的数字系统没有位值制，只有七个基本符号。古玛雅人使用20进位，古巴比伦人使用60进位，与算筹记数法相比，三者在记数和运算中更加繁琐、复杂。中国是世界上最早采用十进位值制的国家，比古埃及早1000多年。

春秋战国时期，中国古人不但发明了十进位值制，还发明了九九乘法表。九九乘法表，又称"九九

in vertical direction, the tens digit was in horizontal direction and the hundreds digit was again in vertical direction, so on and so forth. Zero was left a blank.

This Chinese ancient decimal counting rods calculation is a great invention to the world's history of mathematics. Compared with other ancient civilizations, the rod arithmetic method has a dominant advantage. The number system of ancient Rome was only composed of seven basic symbols instead of operating in different digit places. The ancient Maya employed

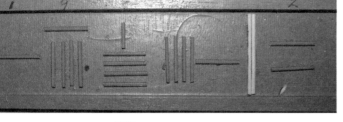

● 算筹
Counting Rods

表""小九九",是中国古代筹算中进行乘法、除法、开方等运算的基本计算规则。后来,九九乘法表向东传入朝鲜、日本,又经过丝绸之路向西传入印度等国,直到13世纪欧洲才开始使用九九乘法表。

the binary-decimal conversion. And the ancient Babylonian used the sexagesimal system. All of the three counting methods are more complicated than the rod arithmetic. China is the earliest country to apply the decimal system, which is more advanced than the ancient Egyptian in more than 1,000 years.

In the Spring and Autumn Period and Warring States Period, the ancient Chinese not only invented the decimal system but also created the multiplication table which is also called nine-times table. With a history of more than 2,000 years, it is the basic calculation rule for the operations of multiplication, division and radication in rod arithmetic. Later, the multiplication table was introduced eastward into North Korea, Japan and westward along the Silk Road into India and other countries. The European didn't employ the multiplication table until 13th century.

> 算盘与珠算

在中国民间传说中,算盘是远古时代黄帝手下一个叫隶首的人发明的。由于人们在日常生产和生活中都需要计算,却没有方便的计算

> Abacus and Abacus Arithmetic

According to the legend, abacus was created by a servant of the Emperor Huang, Li Shou in ancient China. Due to the absence of a convenient tool for daily

- 算盘

算盘一般为长方形、木制,框架里固定着一根根小木棍,木棍上穿着木珠,中间的横梁把算盘分成上、下两部分。

Abacus

The abacus is mainly with a rectangle wooden frame to which wood beads strung with rows of wood sticks are fixed. And a horizontal beam divides the frame into two parts.

工具，隶首就想出了一个办法：到河滩捡回不同颜色的石片，给每块石片都打上眼，用细绳逐个穿起来。每10个或100个石片中间穿一个不同颜色的石片，这样清算起来就省事多了。此后，他又想到一个更加实用的办法，把石片每10颗一穿，穿成100个数，放在一个大泥盘上，在上边写清数位，如十位、百位、千位、万位。这就是算盘的雏形。

算盘是在算筹的基础上发展起来的。由于算筹体积大，携带不方便，且无法满足大量、复杂的数字计算，因此被制作更加简单、运算方式更加快捷的算盘取而代之。明代典籍《鲁班木经》中有关于算盘规格的详细记载："算盘式：一尺二寸长，四寸二分大。框六分厚，九分大……线上二子，一寸一分；线下五子，三寸一分。长短大小，看子而做。"明代算盘的结构已与现在无异。算盘由外框、算珠、横梁三部分组成，有木头、金属、玉等不同材质。算盘一般为长方形框架，由一根横梁分为上、下两部分，算珠用小木棍（称作"档"）

calculation, Li Shou found a method: he collected stone chips of different colors along the river shoal and strung them together with thin thread. Between each ten or hundred chips, he added a stone chip of different color. As thus, counting became more easy and convenient. Then, he came up a more practical approach which was to strung ten chips together as one set and made it to a hundred sets and put them on a huge clay plate marked with digits such as the tens, the hundreds, and the thousands, which was the primitive abacus.

Abacus was considered to be developed on the basis of counting rods. Owing to the inconvenience rendered by its huge volume and the inadequateness in complicated calculation, the counting rods were replaced by the abacus which can be easily produced and operated. According to the classic of the Ming Dynasty (1368-1644), *Carpentry Classic of Lu Ban*, it has detailed records about the size of abacus, quoted as: "The abacus, with a length of 40 cm and a width of 14 cm; the frame, with a thickness of 1.98 cm and a width of 2.97 cm… There are two beads above the beam which cost a length of 3.67

- 《清明上河图》【局部】（北宋 张择端）

  据专家考证，在《清明上河图》中的一家名为"赵太丞家"的药铺桌子上放有一架算盘。

  Life along the Bian River at the Pure Brightness Festival, by Zhang Zeduan (Part) (Northern Song Dynasty, 960-1127)

  According to the experts, there is an abacus placed on the front desk of a herbal medicine store with a signboard inscribed with its name as *Zhao Tai Cheng Jia* in this painting.

固定在框架上。算盘分为二五珠式和一四珠式两种，明代时一四珠式算盘使用较为广泛。二五珠式算盘，横梁上部有两颗算珠（一颗算珠代表五个数），横梁下部有五颗算珠（一颗算珠代表一个数），因一斤为十六两，所以采用十六进制

cm and five beads beam the line which cost a length of 10.33 cm. Basically, the size of the abacus is fixed in accordance of the size and number of the beads." The structure of the abacus in the Ming Dynasty is almost the same with the modern one. The abacus is composed by the frame, beads and beam, which

法进行计算。一四珠式算盘，横梁上部有一颗算珠（一颗算珠代表五个数），横梁下部有四颗算珠（一颗算珠代表一个数），改十六进制为十进制进行计算。

算盘的运算简便，明代以后，以算盘为工具的计算方法珠算逐渐代替了筹算。"珠算"一词最早出现在汉代徐岳撰写的《数术记遗》一书中："珠算，

is usually made from wood, metal, jade and other materials. It is mainly with a rectangle frame and divided into two parts with a beam. The beads are strung and fixed by thin sticks (also called *Dang*) to the frame. It has two types: the two-five style and the one-four style. In the Ming Dynasty, the one-four style was wildly populor. The two-five style had two beads (each bead represents five numbers) above the beam and five beads (each bead represents one number) below it. In order to follow the production and living habits, so the abacus applied the hexadecimal system. The one-four style had one bead (each bead represents five numbers) above the beam and four beads (each bead represents one number) below it and applied decimal system.

Because of its convenience, after the Ming Dynasty, the abacus gradually replaced the counting rods and became the major calculation tool. The Chinese word of abacus, *Zhu Suan*, firstly appeared in the book *Notes of Mathematics* written by Xu Yue in the Han Dynasty (206 B.C.-220 A.D.), quoted as: "The abacus, counts with four beads in three ways." Aside from the basic

• 算盘在当代的应用（图片提供：FOTOE）
Modern Application of Abacus

控带四时，经纬三才。"珠算不仅能够进行加、减、乘、除等基本运算，还能计算房屋、土地的面积。明代商人程大位集各家所长编写了《直指算法统宗》一书，该书对普及珠算起了巨大作用。从明代开始，珠算极为盛行，先后传到日本、朝鲜和东南亚各国。近些年，珠算在美洲也逐渐流行起来。

operations of addition, subtraction, multiplication and division, the abacus also can calculate the area of house or land. Cheng Dawei, a businessman in the Ming Dynasty, wrote the book *Instruction of Abacus Arithmetic*, which has exerted a great influence on the popularization of abacus. From the Ming Dynasty, the abacus prevailed across the country and then was introduced to Japan, Northern Korea and the countries in the Southeast Asia. Lately, it also starts to catch on in America.

> 古代数学著作

从春秋时期开始，中国古代数学不断发展，出现了许多优秀的数学家，他们结合自身研究和历史经验撰写了一批重要的数学著作。

刘徽与《九章算术注》

《九章算术》是中国历史上第一部系统完整的数学著作，成书于西汉时期（前206—公元25）。《九章算术》被后世奉为十大算学经典之首，它对中国数学体系的影响与古希腊数学著作《几何原本》对西方数学体系的影响不相上下。

东汉时，《九章算术》中的一些数学算法已逐渐落后，数学家刘徽为其重新注释，并添加了更为合理、科学的数学理论和计算方法。他用"率"来命名方程（即现代数

> Ancient Works on Mathematics

From the Spring and Autumn Period (770 B.C.-476 B.C.), with the continuous development of mathematics in ancient China, many excellent mathematicians appeared. And based on their studies and experiences, they wrote several classic books of great significance.

Liu Hui and *Commentaries of Nine Chapters on the Mathematical Art*

Nine Chapters on the Mathematical Art, finished in the Western Han Dynasty (206 B.C.-25 A.D.), is the first book systematically giving an introduction on mathematics. It is considered as the top masterpiece among the Ten Great Books on Mathematics and has exerted equal influence on the western mathematics' development with the book *Euclid's Elements* of ancient Greek.

刘徽像
Portrait of Liu Hui

In the Eastern Han Dynasty (25-220), several arithmetic operations gradually became laggard. So another mathematician Liu Hui added new commentaries for this book, which put some more reasonable and scientific mathematical theory and calculation methods. He named the augmented matrix after the Chinese character *Lv*(ratio), determined three basic operations of multiplication all over, common division, and reduction of fractions to a common denominator, used the similarity and difference of the number to elaborate the operation of common denominator reduction, fraction reduction and four arithmetic operations and the rules of simplification, and noticed the non-repeating infinite characteristics of the irrational root and used decimal fraction to approach the result of irrational root. In the *Commentaries of Nine Chapters on the Mathematical Art*, Liu Hui also demonstrated the calculation principle of the Gougu Theorem (Pythagorean Theorem) and established a resemblant theory and developed the relevant measure method and also formed a similar theory with Chinese

学中线性方程组的增广矩阵），确定了遍乘、通约、齐同等三种基本运算，用数的同类与异类来阐述通分、约分、四则的运算以及繁分化简等运算法则。他还认识到无理方根开方不尽的现象，用十进分数无限逼近无理根的方法进行运算。在《九章算术注》中，刘徽还论证了有关勾股定理的计算原理，建立了相似勾股形理论，发展了勾股测量术，形成了中国特色的相似理论。他还用割圆术这一当时最为科学的计算方法，证明了圆面积的精确公

九章

《九章算术》是中国最重要的一部经典数学著作,它不仅在中国数学史上具有重要地位,同时也对世界数学的发展作出了杰出的贡献。《九章算术》中的九章分别指:

方田:田亩面积计算。提出了各种多边形、圆形、弓形等的面积公式。

粟米:谷物粮食的按比例折换。提出比例算法。

衰分:比例分配算法。

少广:已知面积、体积,反求其一边长和径长等。介绍了开平方、开立方的方法。

商功:土石工程、体积计算。给出了各种立体体积公式和工程分配方法。

均输:合理摊派赋税。用衰分术解决赋役的合理负担问题。

盈不足:双设法问题。提出了盈不足、盈适足和不足适足、两盈和两不足三种类型的问题。

方程:一次方程组问题。采用分离系数的方法表示线性方程组。

勾股:利用勾股定理求解的各种问题。

Nine Chapters

Nine Chapters on the Mathematical Art, is a classic book on mathematics in China. Not only does it take a significant position in Chinese mathematics history, it also achieved a great contribution to the development of mathematics in the world. The nine chapters referred in the title indicate:

Land Survey: it focuses on the area calculation of land and records the calculation formula of the areas of polygon, circle and segment of a circle.

Grain Conversion: it focuses on the conversion of various grains and raises proportion arithmetic.

Distribution: it focuses on the matters of prorated distribution.

Length Calculation: With the area and volume, it teaches how to calculate the length of a side or diameter, and introduces the calculation method of extraction of square and cubic roots.

Volume Calculation: it gives various calculation methods referring to the volume calculation in civil engineering and distribution.

Distribution of Taxation: it focuses on finding an appropriate apportion design and uses distribution method to ease the burden of taxes and corvee.

Surplus and Shortage: it is about the questions with two unknowns and raises three scenarios: 1. surplus and shortage; 2. surplus and equal or shortage and equal; 3. surplus and surplus or shortage and shortage.

Equation: it's about linear equation set and applies the method of coefficient separation to indicate system of linear equations.

Gougu Theorem: it applies the Gougu Theorem (Pythagorean Theorem) to solve different problems.

式，得出π接近于3.1416的结论。人们为了纪念他的贡献，把用割圆术法得出的圆周率称为"徽率"。

祖冲之与《缀术》

祖冲之（429—500）是南北朝时期著名数学家、科学家。《缀

characteristics. He applied the most scientific method, cyclotomic method to deduce the accurate formula of the area of a circle and get a conclusion of π approximately equals 3.1416. In honor of Lui Hui and his contribution, people named π after his name, *Hui Lv*.

Zu Chongzhi and *Zhui Shu*

Zu Chongzhi (429-500), is a prestigious mathematician and scientist in the Southern and Northern dynasties (420-589). *Zhui Shu*, is a classic written by Zu Chongzhi, which represents the highest level in mathematics at that time. In the Tang Dynasty (618-907), this book was

• 祖冲之像
Portrait of Zu Chongzhi

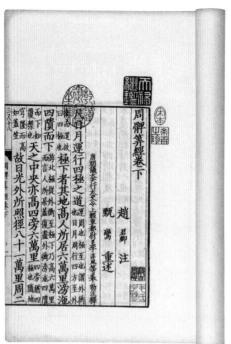

《周髀算经》（民国 影印本）
Mathematical Classic of Zhou Bi (Minguo Period, 1912-1949, Photocopy)

术》是祖冲之的数学著作，代表了当时数学领域的最高水平。在唐代，《缀术》被列入国子监数学教科书——《算经十书》之中。

祖冲之在前人成就的基础上，通过割圆术的方法，得出圆周率的数值在3.1415926和3.1415927之间，这意味着圆周率的数值被精确到了小数点后第7位。祖冲之成为世界上第一位将圆周率值计算到小

listed in the textbook on mathematics, *Ten Computational Canons*, by the Imperial College.

Based on the previous achievements accomplished by ancestors, Zu Chongzhi calculated the value of π is between 3.1415926 and 3.1415927 by cyclotomic method, which means the accuracy of π had been pushed further to the seventh decimal place. Zu Chongzhi also became the first scientist who calculated the value of π to the seventh decimal place, which was 1,100 years in advanced of European scientists. In honor of Zu Chongzhi and his accomplishment, his result of π is called *Zu Lv*.

Zu Chongzhi's study on the value of π satisfied the demand of productive labor at that time. People referred *Zu Lv*, while producing measuring vessels.

算经十书

隋代（581—618）开始在京都设立中央学府——国子监，它是古代教育体系中的最高学府。唐代时，国子监设立了算学馆，将数学列入学生学习的科目。公元656年，《周髀算经》《九章算术》《孙子算经》《五曹算经》《夏侯阳算经》《张丘建算经》《海岛算经》《五经算术》《缀术》《辑古算经》十部汉代至唐代期间的数学经典著作被列入国子监算学教科书，后世称之为"算经十书"。

Ten Computational Canons

Since the Sui Dynasty (581-618), all the authorities afterwards established their central academy, Imperial College, in the capital cities. It was the highest institute in ancient education system. In the Tang Dynasty (618-907), the Imperial College set up the department of mathematics and put mathematus into the required courses. In 656, ten classic books on mathematics from the Han Dynasty to the Tang Dynasty, including *Mathematical*

- 北京国子监古建筑（图片提供：全景正片）
Old Buildings of the Imperial Academy in Beijing

Classic of Zhou Bi, *Nine Chapters on the Mathematical Art*, *Mathematical Classic of Sun Zi*, *Mathematical Classic of Five Administrators*, *Mathematical Classic of Xiahou Yang*, *Mathematical Classic of Zhang Qiujian*, *Mathematical Classic of Sea Islands*, *Mathematical Notes to Five Classics*, *Zhui Shu*, and *Collection of Mathematical Classic*, were compiled to be the textbook in the Imperial College, which was called Ten Computational Canons by later generations.

数点后第7位的科学家，比欧洲人早了1100多年。人们为了纪念他的贡献，把他的计算结果命名为"祖率"。

祖冲之在圆周率方面的研究，适应了当时生产劳动的需要，人们在制造量器时就采用了祖冲之的"祖率"数值。

祖冲之还是一位出色的天文学家，他创立了历法《大明历》，首次运用了岁差概念，使历法的计算更加精确。

Zu Chongzhi was an outstanding astronomer as well and established the *Daming Calendar*, and firstly applied the theory of precession to make the calendar more accurate.

医药学
Medicine

在春秋战国时期，中国的医药学理论就已经基本形成，出现了经络学说和藏象学说等中医理论，确立了"望、闻、问、切"（即四诊法）的诊断方法以及砭石、针刺、艾灸、汤药四大治疗方法。在两千多年的发展过程中，出现了诸多医学典籍，它们至今仍然发挥着重要作用。

In the Spring and Autumn Period (770 B.C.–476 B.C.) and Warring States Period (475 B.C.–221 B.C.), the theory of traditional Chinese medical science already established, including the Theory of Visceral Manifestation Theory and the Meridian and Collateral Theory. It founded the four main diagnosis methods, i.e. observation, auscultation and olfaction, inquiry and palpation, and the four therapeutic methods including *Bian*-stone (a small cone or wedge stone implement formed through stone grinding, used by the ancient Chinese to remit pain by massage), acupuncture, moxibustion and decoction. After a development stretching over two thousand years, it appeared numerous medical canons which still inspire people with great significance.

> 中医理论

中医以"天人合一"哲学观点和辨证论治为基础和核心，同时结合医者以往的医疗经验。中医认为，人的生老病死等生命活动都和自然界有着密切的关联，中医在诊

> Theories of Traditional Chinese Medicine

The theory of traditional Chinese medicine is rooted in the continuous summary of medical experience, which believes that man and the nature are one unity, namely "harmony between the nature and man". According to the theory, there exists close connection between the natural world and law of human activities and the cause of disease. Therefore, doctors of traditional Chinese medicine usually apply different treatment methods even to the same disease and never deal with the nidus solely. The theory of traditional Chinese medicine include the Theory of Viscera-state doctrine, doctrine of main and collateral channels, etc.

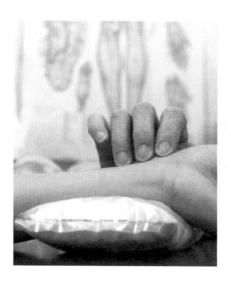

• 切诊（图片提供：全景正片）
Palpation

四诊法

依据中医学的理论体系，中医通过对病人外在症状的观察，来判断病人内在病症，这种诊断方法被称作"四诊法"，即望、闻、问、切。

望诊在中医的诊断技术中占有很重要的地位。望诊的内容主要是观察人的神、色、形、态，以推断体内的变化，其中对面部和舌部的望诊是最常用也是最重要的手段。

闻诊的"闻"，不仅指嗅，也有听的意思。医生依据病人发出的各种声音，如高低、缓急、强弱、清浊来测知疾病的性质。例如，语声重浊，可能是外感风寒、肺气不宣、气郁津凝；疾病末期，病人出现神志不清、语声低微，这是久病正衰、心气虚损、精神散乱的症状。

问诊是对病人或陪诊者进行系统而有目的的询问，询问内容包括病人的体质、生活习惯、起病原因、现在的症状及过去的病史、有无家族病史等。问诊越详细越有利于疾病的诊断与治疗。

切诊包括诊脉以及对病人四肢、胸腹、腰背等部位的触摸按压。后者相当于西医的"触诊"。医书中说"心中了了，指下难明"，足见其神奇之处。总体来说，正常人的脉象应该是快慢适中、不浮不沉、不大不小、从容和缓、节律一致，并能随生理活动和气候环境变化而有相应的变化。

Four Main Diagnosis Methods

According to the theoretical system of traditional Chinese medicine, doctors can tell the interior disease by observing patients' external physical symptoms, which is called "four main diagnosis methods", i.e. observation, auscultation and olfaction, inquiry and palpation.

Observation is of great significance. It mainly refers to deducing the changes inside the body by observing the spirit, color, physique and posture of people, thereinto, the most essential and frequently used methods are facial observation and tongue observation.

Auscultation and olfaction also include listening except smelling. A doctor confirms the nature of the disease by hearing the voice the patient makes: high or low, fast or slow, powerful or weak, clear or turbid. If in a turbid voice, he might have caught a cold and the *Qi* inside the lung cannot go out, leading to the blocking of *Qi* and saliva; if at the advanced stage of a disease, the patient shows the symptoms of obnubilation and low voice, it indicates that he has been suffering from illness for a long time, and with very weak physical condition, heart *Qi*, and mental state.

Inquiry refers to the process in which a doctor asks a patient or his accompany some systematic questions with purpose. The questions consist of the physique, living habits, cause of disease, current symptoms, medical history, family medical history, etc. A detailed inquiry is very helpful to the diagnosis and treatment of disease.

Palpation includes feeling the pulse and touching or pressing the patient's limbs, chest, abdomen, waist, back and some other parts. The latter one is equivalent to the "palpation" in western medicine. The miracle part of this diagnosis can be reflected in the content of the medical book, quoted as: "It is difficult to feel the pulse in practice in spite of a clear mind of theories." Generally speaking, the normal pulse tends to be neither floating nor sunken and neither fast nor slow, moderate in size, gentle in sensation and regular in beating and varies with physical activities and environmental changes.

疗疾病时，根据具体情况采用不同的治疗方法，他们从不孤立看待任何一种生理或病理现象。中医理论包括藏象学说、经络学说等内容。

藏象学说

藏，即脏，是指人体内的脏腑，包括心、肝、肺、脾、肾、胃、肠、胆等；象，是指脏腑的功能活动、病理变化等在人体表面的种种征象。藏象学说是研究人体脏腑的生理功能、病理变化及其相互关系的中医学说，是一种独特的生理病理学理论体系。中医通过外在的"象"，就可以判断出内在的"藏"的各种生理、病理状态，以此为基础来治疗疾病。

Visceral Manifestation Theory

Viscera refer to viscera inside the human body, including the heart, liver, lung, spleen, kidney, stomach, bowel, bladder and so on; Manifestation refers to the external physiological and pathological signs of viscera. Visceral manifestation theory refers to the study of human organs and physiological function, pathological changes and their mutuality,is the theoretical system of a kind of unique physiological pathology. Doctor of traditional Chinese medicine senses the physiological and pathological status of internal viscera through the changes of the external manifestation.With this, traditional Chinese medicine will cure it by adjusting the relationships between viscera.

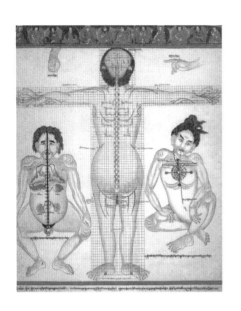

• 中国古画中的《人体内脏图》
Five Human Zang-viscera Chart in the Ancient Chinese Painting

经络学说

经络学说是指研究人体经络的生理功能、病理变化及其与体内脏腑相互关系的学说，包括沟通人体内外、联系五脏的经脉和络脉两大部分。中草药和针灸等治疗手段的使用都要以经络学说为依据。中医认为，身体出现病症，可以通过经络来进行治疗。例如，手阳明大肠经即大肠经，是人体主要经脉之一，疏通大肠经能够有效地防治皮肤病。

Meridian and Collateral Theory

Meridian and Collateral Theory refers to the study on main and collateral channels' physiological function and pathological changes in human body as well as their relationship with viscera. The use of herbs and acupuncture-moxibustion therapy should require the instruction of Meridian and Collateral Theory. Doctors of traditional Chinese medicine believe that the problems of the body can be nursed and treated through the meridian and collateral system. For example, dredging large intestinal meridian, can effectively prevent skin diseases.

针灸

针灸是针与灸的合称,在经络学说的指导下,通过刺激经络上的穴位,使气运行正常,从而达到治疗的目的。针、灸两种方法可同时使用。

针的前身是砭石。砭石是一种经过磨砺而成的锥形或楔形的小石器,古人用其按摩以缓解疼痛。后来砭石发展为石针、骨针。随着冶金术的发明,针具也得到了不断改进,出现了铜针、铁针、金针、银针。针的长短形状也各不相同,《黄帝内经》中记载有九种针具,即镵针、圆针、鍉针、锋针、铍针、圆利针、毫针、长针和大针。九种针的形状、用途各异,医者根据情况选择使用。以针刺激体表的穴位,可以起到治病的作用。

灸法的产生是在火的使用以后。最初是用各种树枝点燃来灸,后来发展为用艾草施灸。艾灸就是将用艾草加工而成的艾绒、艾条或艾炷点燃,熏烤人体穴位,产生温热的刺痛来缓解病痛。艾灸在中国应用得很早,许多医学著作里都有记载。

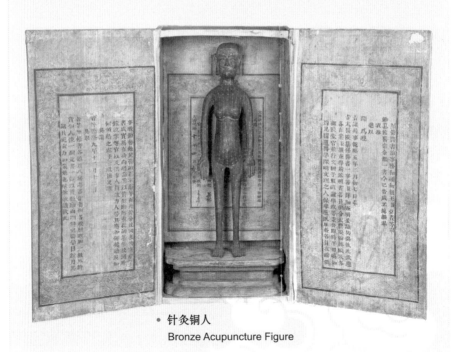

● 针灸铜人
Bronze Acupuncture Figure

Acupuncture-moxibustion Therapy

The acupuncture-moxibustion therapy is a combination of acupuncture and moxibustion. It is a traditional therapy operated under the instruction of the Meridian and Collateral Theory. Through stimulating the acupoints, it can channel the *Qi* in formal movement, thereby curing diseases. Acupuncture and moxibustion can be performed at the same time.

The predecessor of the acupuncture needle was *Bian*-stone which was a small cone or wedge stone implement formed through stone grinding, used by the ancient Chinese to remit pain by massage. Later, *Bian*-stone was developed into stone needles and bone needles. Due to the invention of metallurgy, needle instruments have been improved correspondingly; copper needles, iron needles, gold needles, and sliver needles were also invented. The needles vary in length and shape. *Inner Canon of the Emperor Huang* records nine different needles, including the chisel needle, round needle, arrowhead needle, ensiform needle, beryllium needle, round sharp needle, filiform needle, long needle and large needle. They were different from each other in both shape and usage, and should be used based on the specific condition to stimulate the corresponding acupoints in order to achieve the best treatment effect.

The moxibustion therapy emerged after people knew how to use fire. Initially, people did moxibustion with various lit branches, and then, wormwood. Moxibustion is to light the moxa wool, moxa stick or moxa cone made from wormwood to fumigate and roast acupuncture points of the human body for the purposes of relieving pain and making people feel comfortable. Moxibustion has been used in China for a very long time and recorded in many medical publications.

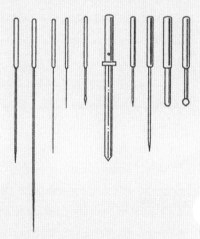

• 不同形质的针
Needles with Different Appearances

• 艾灸
Moxibustion

> 中药学

中药是中医为患者治疗疾病的重要媒介。中医利用药物本身所具有的特性，使人体达到平衡。经过数千年的发展，人们逐渐掌握了药物的使用规律和方法，逐渐形成了一门学科——中药学。

在中医中，动物、植物、矿物等都可入药，其中以植物类最为普遍，有"百草"之称。汉代《淮南子·修务训》一书中有关于神农氏尝百草的记载。神农氏是远古时代的部落首领，相传是中国古代最早认识到植物药性的人。

不同的中药具有寒、热、温、凉四种不同的特性，还有些中药性质平和，既不过热，也不过寒，即"平性"。寒、热、温、凉、平五种不同的特性被称为中药的

> Science of Chinese Pharmacology

Traditional herbal medicine is an important approach to cure diseases. It can maintain an internal balance by taking medicines with various properties and flavors. With a development of thousands of years, people gradually acquired the dosage pattern and application method of the medicine and then developed it into a major subject, science of Chinese pharmacology.

According to traditional Chinese medicine, all the animals, plants and minerals can be deployed into the medicine, especially the plant, also called "hundred herbs". In the book *Huai Nan Zi: Cultivation*, it has the record about the legend of Shennong tasting hundreds of herbs. Shennong was a tribal chief in ancient China and also was the first man who learnt the drug property of plants.

• 铜制药杵臼（明）
药杵臼是一种常见的中药加工工具，最迟在战国时期就已出现。

Copper Medicinal Mortar and Pestle (Ming Dynasty, 1368-1644)
They are common tools used for pharmaceutical production. It appeared no later than the Warring States Period (475 B.C.-221 B.C.).

"气"。中医利用中药不同的特性来调节人体的寒热失衡情况以达缓解并治愈疾病的目的。例如寒凉的中药主要用于治疗热症，像黄连、黄芩偏于苦燥，能够清热化湿。温热的中药主要用来治疗寒症，像吴茱萸温胃散寒，能够治疗胃痛酸吐、痛经等。除了具有五种"气"之外，中药还具有五种味，分别是酸、苦、甘、辛、咸。例如人们吃

Different herbal medicines possess various properties, including cold, hot, mild, cool and neutral which are called the *Qi* of herbs in traditional Chinese medicine. Different properties can be used to adjust human body's unbalanced state of cold and hot to remit or cure the disease. For example, the herbal medicine with cold property mainly targets the hot symptoms, such as coptis chinensis and scutellaria baicalensis with bitter and arid quality can clear away the hot and damp. And the herbal medicine with warm or hot property is mainly used to deal with the cold symptoms, such as fructus evodiae can warm stomach and clear away the cold *Qi*, which can deal with stomachache and dysmenorrhea. Aside from the five *Qi*, herbal medicine also possesses five flavors, namely sour, sweet, bitter, pungent and salty. For example, if people eat mustard they will feel the pungent flavor and get a relaxed comfort, which indicates that spicy food has the function of removing obstructions. Through the long-term practice, people have realized that among all traditional Chinese medicines, sour medicines have the properties of astringent and blocking up; bitter medicines, discharging fire and eliminating dampness; sweet medicines,

了芥末，会有辛辣、通窍的感觉，这是因为辛味具有开通、发散的作用。在长期的实践中，人们还慢慢总结出：酸味药能够收敛、涩滞，苦味药能够泻火、燥湿，甘味药能够补益、和缓，咸味药能够泻下、软坚，淡味药能够利水渗湿。利用中药这些特性，可达到恢复人体平衡的目的。

《神农本草经》中记载："药有阴阳配合……有单行者，有相须者，有相使者，有相畏者，有相恶者，有相反者，有相杀者，凡此七情，合和视之。""单行""相须""相使""相畏""相杀""相恶"和"相反"是指中药之间的七种关系，称为"七情"。中医在配制中药药方时，需注意药物之间的关系，以保证药性相配，达到治愈疾病的目的。患者在服用中药时，同样需要注意药物之间的关系，否则会产生不良反应，甚至中毒，同时，还要注意食物上的禁忌，例如服用人参不宜吃萝卜、服用茯苓不宜吃醋等。

tonifying and recuperating body mildly; salty medicines, purgation and softening; flat medicines, disinhibiting water and clearing dampness. These properties of herbal medicines can help people recover the inner balance of human body.

According to *Shennong's Herbal Classic*, "Herbal medicine emphasizes the coordination of *Yin* and *Yang*, such as the single, the matching, the assisting, the restrained, the invalid, the contrary and the dispelling which should be used or deployed under an overall consideration." These seven relationships between herbal medicines are called Seven Affections. The prescriptions should refer different relationships between the medicines to assure the drug properties and cure the disease. And the patient also needs to pay attention to the relationships. Otherwise, it will cause several side effects or even toxic reaction. Moreover, the taboo on the combination of food requires special attention as well. For example, it is inappropriate to combine ginseng with radish, as well as the tuckahoe and vinegar.

- 多种多样的中草药（图片提供：图虫创意）
  Various Traditional Herbal Medicines

神医华佗

华佗是东汉末年著名医学家,人们经常用"华佗再世"来形容医术高明的人。华佗是最早采用外科手术进行治疗的医学家,被后世尊为"外科鼻祖"。他还发明了世界上最早的麻醉药——麻沸散,利用麻沸散对患者施行全身麻醉后进行外科手术。唐代医学家孙思邈编写的《华佗神方》一书中对麻沸散有详细的记载。华佗不仅在外科方面成就卓著,还模仿虎、鹿、熊、猿、鸟的动作创造了五禽戏。五禽戏是一套能够舒展筋骨、疏通经脉的医疗体操,对强身健体、防治疾病有很好的效果。

Hua Tuo

Being an outstanding doctor in the Eastern Han Dynasty (25-220), Hua Tuo's name is usually borrowed to praise skillful doctors. He is the first doctor who applied surgical operation in the treatment, so Hua Tuo is also praised as "the founder of China's surgery". He developed the first narcotic in the world, Powder for Anesthesia and conducted surgical operation after the general anesthesia, which had been recorded with details in the book *Prescriptions of Hua Tuo*, written by the famous doctor in the Tang Dynasty (618-907), Sun Simiao. Aside from his achievement on surgery, he also created the famous Five-animal Exercises, a physical exercise taken by imitating the actions of tiger, deer, bear, monkey and crane. It is a therapeutic exercise which can stretch muscles and bones, clear the passages of meridians and collaterals, build up strength and resist or cure diseases.

● 华佗像
Portrait of Hua Tuo

> 古代医学典籍

中医理论和治疗方法的传承和发展要靠从事医疗工作的人来完成。古代，众多的医学家为此付出了很大的努力，为后世留下了许多具有参考价值的医学典籍。

《黄帝内经》

《黄帝内经》成书于春秋战国时期，是中国最早的医学理论著作，是对上古时期医学经验的总结和归纳。黄帝，又称轩辕氏，是中国远古时期的部落联盟首领。传说，黄帝是上古时期医学的开创者之一，他命精通医术的岐伯撰写医书，为医学发展奠定了基础。后人为了纪念二者的贡献，又把中医称为"岐黄之术"。但《黄帝内经》并非黄帝

> Ancient Canons of Traditional Chinese Medicine

The inheritance and development of Chinese traditional medicine theories and treatments mostly rely on the experts who are engaged in this subject. In ancient time, many medical scientists had exerted

• 黄帝像
Portrait of the Emperor Huang

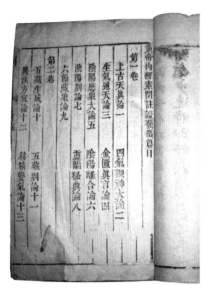

Photocopy of *Inner Canon of the Emperor Huang*

their effort in this area and left numerous medical canons of high reference value.

Inner Canon of the Emperor Huang

Inner Canon of the Emperor Huang, finished in the Spring and Autumn Period and Warring States Period, is the earliest medical canon in China, which concludes the medical experience in ancient times. The Emperor Huang was a tribal chief in ancient China. Legend has it that the Emperor Huang was one of the inaugurators who founded traditional Chinese medicine. He asked the skillful doctor Qi Bo to write a medical book which laid the foundation of the development of traditional Chinese medicine. So people in later generations often call this science as the Art of *Qi* and Huang in order to commemorate both of them. However, *Inner Canon of the Emperor Huang* was not written by the Emperor Huang but was compiled in the Emperor Huang's name. The book has two parts: *Plain Questions*, and *Therapy of Acupuncture*, and states the theories of human physiology, human anatomy, internal organs, meridians and collaterals, cause of disease, pathogenesis, symptoms, diagnostic method, treatment principle, health preserving and

所著，而是后人托黄帝之名编写的。全书分为《素问》和《灵枢》两部分，采用黄帝提问、岐伯回答的对话方式讲述了人体生理、解剖、脏腑、经络、病因、病理、病征、诊法、治疗原则、养生以及针灸等医学内容。除此之外，《黄帝内经》还包含天文学、地理学、心理学、社会学、哲学等方面的内容，堪称百科全书。

《难经》

春秋时期，有位名叫秦越人的名医，医术十分高明，人们用黄帝时期一位医术高超的神人"扁鹊"的名号来称呼他。扁鹊（秦越人）创造了"四诊法"——望、闻、问、切，奠定了中医的诊断与治疗

- 扁鹊（秦越人）像

扁鹊即秦越人，他曾经通过观察蔡桓公的外貌、体态准确地判断出其生病的部位，但是蔡桓公却不愿接受治疗，病情逐渐从表皮、皮肤、肌肉加深到骨髓，最后难以医治而死去。

Portrait of Bian Que (Qin Yueren)

Bian Que, originally named Qin Yueren, once performed an accurate diagnosis merely through observing the appearance and figure of Duke Caihuan. But Duke Caihuan refused to accept any medical treatment. So the nidus was gradually aggravated from the exterior skin to muscle, and finally to the marrow. At last, the Duke died from a deteriorated yet uncurable disease.

acupuncture and moxibustion through conversations between the Emperor Huang and Qi Bo. Besides, *Inner Canon of the Emperor Huang* also includes the contents on astronomy, geography, psychology, sociology, and philosophy. So it is considered as an encyclopedia.

Collection of Medical Problems

In the Spring and Autumn Period (770 B.C.-476 B.C.), there was a famous doctor named Qin Yueren. Due to his miraculous medical skill (it says that he could bring the dying back to life), he was honored after the name of a legendary doctor Bian Que in the reign of the Emperor Huang by the public. He created four main diagnosis methods, including observation, auscultation and olfaction, inquiry and palpation, which laid the foundation of the diagnosis and treatment of traditional Chinese medicine.

People in the later generation wrote the medical canon *Collection of Medical Problems* in the name of Bian Que, which was finished no later than the Eastern Han Dynasty (25-220). Through a conversation of interrogation and reply, this book explains 81 medical problems. So it is also called *Collection of Eighty-one Problems*. It mainly focuses on

方法的基础。

后人托扁鹊之名撰写了医学典籍《难经》，此书最迟成书于东汉，以问答的方式解释了81个疑难问题，故又称《八十一难经》。全书所述以基础理论为主，还分析了一些病症。其中一至二十二难主要讲述脉学，二十三至二十九难主要讲述经络，三十至四十七难主要讲述脏腑，四十八至六十一难主要讲述疾病，六十二至六十八难主要讲述腧穴，六十九至八十一难主要讲述针法。

《难经》明确提出"伤寒有五"，即中风、伤寒、湿温、热病、温病五类病症。除此之外，对泻痢类疾病也有所阐述。全书清晰简明，被后人奉为医学经典著作。

《神农本草经》

神农氏，即中国远古时代的部落首领炎帝。相传他遍尝百草，以身试药而死，被后人奉为"医药之

the basic theories, as well as several symptom analyses. 1-22 chapters are about sphygmology; 23-29 chapters are about meridian and collateral; 30-47 chapters are about viscera; 48-61 chapters are about diseases; 62-68 chapters are about acupoints; 69-81 chapters are about acupuncture therapy.

It is clearly presented in the *Collection of Medical Problems* that five symptoms of stroke, typhoid fever, damp and warm syndrome, fever, and febrile disease are concluded as typhia. Besides, it also depicts the symptom of dysentery. With a clear and concise narration, this book is praised as a classic canon by later generations.

● 神农像

Portrait of Shennong

- 《神农本草经》书影
  Photocopy of *Shennong's Herbal Classic*

Shennong's Herbal Classic

Shennong, was the tribal chief in ancient China, the Emperor Yan. Legend has it that he died from tasting the unknown herbs. After his decease, he was praised as the God of Medicine by later generations in China. *Shennong's Herbal Classic*, was finished in the Qin and Han dynasties (221 B.C.-220 A.D.), and was written by doctors at that time in the name of Shennong. It is the earliest herbal canon founded in China.

Shennong's Herbal Classic, with three volumes, includes 365 kinds of medicines (with 252 plants, 67 animals, and 46 minerals) and 13 medical theories, firstly raised the prescription theory of "emperor, official, assistant and guide" (indicating the different functions of four types of herbal medicines) which is still referred by the formulas of Chinese medicine. As to the method on how to deploy different medicines, it concludes seven conditions, namely the single, the matching, the assisting, the restrained, the invalid, the

神"。《神农本草经》是秦汉时期的医学家托神农氏之名而编写的。此书是中国现存最早的一部药物学专著。

《神农本草经》全书共三卷，收载药物三百六十五种，其中植物药二百五十二种，动物药六十七种，矿物药四十六种。该书归纳了十三种医药理论，首次提出了"君臣佐使"（即"君主""臣僚""僚佐""使者"，在中药处

神农尝百草

上古时期，人们常常会因为误食毒草而丧命，生病也不知道该用什么药医治。神农氏决定为民排忧解难，他带领一批臣民进入深山去实地考察。神农氏在山上尝遍了百草，详细地记述了所尝草药的产地、性质、采集时间、入药部位和主治病症等。神农氏还发现谷子、高粱、稻谷等植物种子是可以充饥的，于是将其种子带回去种植，这就是后来的五谷。有一次，在试尝一种草药的时候，神农氏身中剧毒而亡。人们感念神农氏的贡献，尊称其为"医药之神"。

Shennong Tasting Hundreds of Herbs

In the ancient times, people often mistook the poisonous plant and lost their lives, and they also had no idea on the effects of different herbs. Therefore, Shennong decided to solve these problems for the people. He led several officials to the mountains to carry out the investigation. He tasted all the herbs and plants on the mountain and recorded their production area, quality, collecting time, officinal parts and targeting symptoms in detail. He also found that millet, Chinese sorghum and rice are edible. So he took their seeds back to the tribe and started to grow them in large scale. And they were later called five cereals. Once he tasted an unknown plant, he died from toxication. In honor of his accomplishment, Shennong is also called the God of Medicine.

- 《神农采药图》
  Shennong Collecting Herbs

方中起不同作用的各味药）的方剂理论，一直被后世方剂学所沿用。关于药物的配伍情况，书中概括为"单行、相须、相使、相畏、相恶、相反、相杀"七种，指出药物的配伍前提条件：有的药物合用，可以相互加强作用或抑制毒性，宜配合使用；有的药物合用会使原有的药理减弱，这样的药物要避免同时使用。

《神农本草经》对药物性味作了详尽的描述，指出寒、热、温、凉四气和酸、苦、甘、辛、咸五味是药物的基本性情，可针对疾病的寒、热、湿、燥性质不同选择用药。寒病选热药，热病选寒药，湿病选温燥之品，燥病须凉润之流，配伍用药。

《伤寒杂病论》

《伤寒杂病论》是中医药学的经典著作，为东汉末年著名医学家张仲景所著，他被人们称为"医圣"。东汉末年，朝政腐败，民不聊生，疾病流行却得不到有效地控制。由于当时医学不发达，人们大多迷信巫术，很多人因此丧命。张

contrary and the dispelling, which gives the preconditions on the compatibility of medicines. It believes that a harmonious ingredient can enhance the drug properties or restrain the toxicity and inappropriate set can weaken the effect of drugs.

In the *Shennong's Herbal Classic*, detailed depictions regarding to the flavor and quality of the herbal medicines are stated. It points out the four *Qi* (cold, hot, warm and cool) and five flavors (sour, bitter, sweet, pungent and salty) are basic property of herbal medicines. The doctor can write the prescriptions based on different symptoms of diseases (cold, hot, damp and arid). The cold disease can be dealt with hot medicine, hot disease to cold medicine, damp disease to warm and arid medicine and arid disease to cold medicine.

Treatise on Cold Pathogenic and Miscellaneous Diseases

Treatise on Cold Pathogenic and Miscellaneous Diseases, is a classic canon on traditional Chinese medicine. It is written by a famous doctor Zhang Zhongjing in the late Eastern Han Dynasty (25-220), who is also called the Sage of Medicine. In the late Eastern Han Dynasty, gov-

仲景对此十分痛心，建安十年（205年），他开始撰写《伤寒杂病论》一书。

书中共列有300多计药方，其中许多药方经现代医学临床试验证明依然有效，被后人誉为"方中之祖"。张仲景还开创了"辨证论治"的医学理论。辨证论治，是指要运用各种诊断方法，辨别不同的患者症状，根据病人的生理特点，并结合时令节气、地区环境、生活习俗等因素进行分析，研究其致病原因，然后确定恰当的治疗方法。《伤寒杂病论》中还首次记载了人工呼吸、药物灌肠等治疗方法。

ernment was sunk in the corruption and ordinary people lived a difficult life and various diseases prevailed without efficient control. Due to the underdeveloped medical science, most of the people believed in witchcraft which cost them a large amount of fortune or even their lives. Zhang Zhongjing grieved over such painful situation and in the year of 205, he started to write the book, *Treatise on Cold Pathogenic and Miscellaneous Diseases*.

There lists more than 300 prescriptions. Many of them are proved efficient through clinical tests of modern medicine. So the book is also praised as the Ancestor of Prescription Books. Zhang

• 张仲景像
Portrait of Zhang Zhongjing

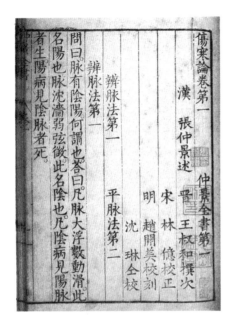

• 《伤寒杂病论》书影
Photocopy of *Treatise on Cold Pathogenic and Miscellaneous Diseases*

Zhongjing also developed the medical theory of syndrome differentiation and treatment which indicates that the doctor should find out the cause of the disease and determine an appropriate treatment plan based on the physiological characters of the patient and other elements such as seasons, local environment, and customs, through different diagnosis methods and various analyses on different symptoms. In the *Treatise on Cold Pathogenic and Miscellaneous Diseases*, it firstly included artificial respiration, and medical enema, etc.

Collection of Invaluable Prescriptions

Sun Simiao (581-682), is a medical scientist and pharmacologist in the Tang Dynasty (618-907), who is called the King of Medicine by later generations. *Collection of Invaluable Prescriptions*, is an important medical canon written by Sun Simiao. With its full name as

《千金方》

孙思邈（581—682）是唐代医学家、药物学家，被后人称为"药王"。《千金方》是孙思邈最为重要的医学著作，全称为《备急千金要方》，被誉为中国最早的一部临床医学百科全书。他认为"人命至重，有贵千金"，故以此为名。该书共三十卷，第一卷为医学总论及本草等内容，第二至四卷论述妇科病，第五卷论述儿科病，第六卷论述七窍病，第七至十卷论述诸风

● 孙思邈像
Portrait of Sun Simiao

（风寒、风热、风湿等病症的总称）、脚气、伤寒等，第十一至二十卷按脏腑顺序介绍了一些内科杂病等，第二十一卷论述消渴、淋闭等症，第二十二卷论述疔肿痈疽，第二十三卷论述痔漏，第二十四卷论述解毒及其他杂治，第二十五卷论述备急诸术，第二十六至二十七卷论述食治及养性，第二十八卷论述平脉，第二十九至三十卷论述针灸孔穴。

孙思邈行医一生，以解除病人痛苦为唯一职责，对待病人"皆如至尊"，是中国医德思想的创始

Emergency Collection of Invaluable Prescriptions, it is praised as the earliest encyclopedia on clinical medicine in China. He believed that the life is much more valuable than the gold, hence the name. This book has 30 chapters: chapter 1 is mainly about the general introduction of traditional medicine and herbs; chapter 2-4 are about gynopathy; chapter 5 is about paediatrics; chapter 6 is about the disease of seven apertures; chapter 7-10 are about wind caused diseases (wind chill, wind-heat and wind-damp), dermatophytosis, and typhoid fever; chapter 11-20 state several internal

人，被西方称为"医学论之父"。在行医过程中，孙思邈一直致力于药物的研究，曾经到峨眉山、终南山、太白山等地采集中药，进行临床试验，并系统、全面、具体地论述了各种药物种植、采集、收藏等方面的内容，是第一个将野生药物进行人工种植的医学家。

《本草纲目》

《本草纲目》为明代著名的医学家、药物学家李时珍（1518—1593）所著。他出生在一个世代行医的家庭，从35岁开始，前后历时27年，参考了800多种书籍，经过三次大规模的修改，终于完成了药物学巨著——《本草纲目》。

《本草纲目》共五十二卷，一百九十多万字，共记载了1892种药物，其中有374种为李时珍所加，并附有11000多个药方，还附带了1100多幅药物形态图，这些都源自李时珍亲自考察和实践经验。《本草纲目》吸收了历代本草著作的精华，纠正了以往的错误，并有很多重要发现和突破，是16世纪

diseases in the order of viscera; chapter 21 is about the syndrome of excessive water drinking and prostatitis; chapter 22 is about staphylococcus aureus infection and ulcer; chapter 23 is about anal fistula; chapter 24 is about detoxication and other miscellaneous illnesses; chapter 25 is about several emergency treatments; chapter 26-27 are about dietary therapy and nature cultivation; chapter 28 is about pulse diagnosis; chapter 29-30 are about acupuncture and moxibustion.

Sun Simiao spent all his life in the cause of eliminating patients' suffering. As he treated his patients with respect and concern, he also was the first man who set up the medical ethics in China. He was called the Father of Medicine by the western world. In his medical practice, Sun Simiao was always dedicated to the study on the herbal medicine. He once collected herbs on Mount Emei, Mount Zhongnan and Mount Taibai and also conducted clinical tests and then systematically and specifically explained the planting, collecting and storing of herbal medicines. He was the first medical scientist who planted wild herbs by artificial cultivation.

• 李时珍像
Portrait of Li Shizhen

Compendium of Materia Medica

Compendium of Materia Medica, is written by the prestigious medical scientist and pharmacologist in the Ming Dynasty (1368-1644), Li Shizhen (1518-1593). He was born in a family of doctors. At the age of 35, he started to write this medical canon. In 27 years, He referred to more than 800 books and records and went through three massive modifications and finally finished *Compendium of Materia Medica*.

Compendium of Materia Medica records 1,892 medicines and over 11,000

• 《本草纲目》书影
Photocopy of *Compendium of Materia Medica*

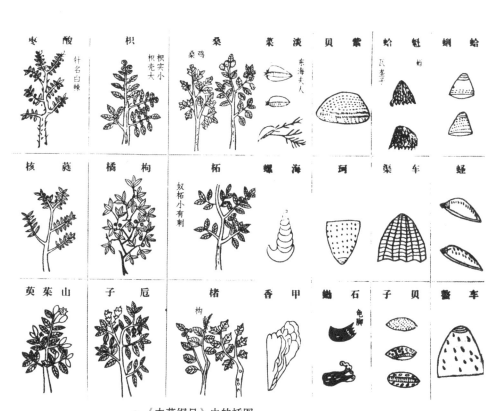

•《本草纲目》中的插图
Illustrations in *Compendium of Materia Medica*

时期中国最系统、最完整、最科学的一部医药学著作，英国生物学家达尔文称赞它是"中国古代的百科全书"。

李时珍把植物分为草部、谷部、菜部、果部、木部五类，这种分类方法在当时是独一无二的。《本草纲目》中详细记述了每一种药物的"释名"（即名字）、"集解"（即解释）、"修治"（即炮制）、"气味""主治""发明"（作者的发现）、"附方"等内容。

prescriptions with 1.9 million Chinese characters in 52 volumes, and contained over 1,100 illustrations of herbs which are referred to his personal field investigations and practices. This book takes over the essences of all the past herbal books, corrects the previous mistakes, and has many important breakthroughs. So it can be considered as the most systematic, intact and scientific medical canon in 16th century in China. The book was praised by the famous British biologist, Darwin as the "encyclopedia of ancient China".

Li Shizhen divided the section of plant medicine into five categories, including chapter of grass, chapter of cereal, chapter of vegetable, chapter of fruit and chapter of wood. This classification method was unprecedented at that time. *Compendium of Materia Medica* records each medicine's name, introduction, production process, smell, major functions and new qualities found by author, as well as several prescriptions.

农学
Agronomy

　　中国是世界上最早发展农业的国家之一，在新石器时代早期就已经出现了人工种植的稻谷。重视农业是古代君主治理国家的基本思想，古代劳动人民也在实践的基础上不断总结耕种的经验。在漫长的历史发展中出现了许多先进的农业生产工具和技艺、杰出的农学家及农业著作以及功绩卓著的水利灌溉工程等。

China is one of the earliest countries that developed agriculture in the world and at the latest, the artificial cultivation of rice appeared in the early phase of the Neolithic Age (approx. 10,000–4,000 years ago). The ancient imperial emperors attached great importance to agriculture and regarded it as the basic concept of governing a nation. At the same time, the ancient Chinese labouring people continually summarized experiences in cultivation based on their agricultural practices, and numerous advanced agricultural tools and techniques, outstanding agriculturists and famous books of agriculture as well as famous and meritorious water conservancy irrigation projects emerged in a long history of development.

> 农业生产与二十四节气

自秦代开始,中国历朝历代都在法律上赋予农业最高的地位。唐太宗曾说过:"凡事皆须务本。国以人为本,人以衣食为本,凡营衣食,以不失时为本。"

农作物的播种、收获等都受到太阳光照的影响,根据季节的变化而变化。早在春秋战国时期,古人就已经确定了夏至、春分、秋分、冬至四个节气,并认识到夏至这天的白日最长、夜晚最短,冬至则是白日最短、夜晚最长,春分、秋分为昼夜等分。秦汉时期,根据太阳在公转轨道上的位置,将一年分割为平均的二十四份,并冠以不同的名字。公元前104年,汉武帝颁布《太初历》,正式把二十四节气列入历法。二十四节气中的"气"是

> Agricultural Production and 24 Solar Terms

Ancient China gave the highest legal position to agriculture in the following successive dynasties from the Qin Dynasty (221 B.C.-206 B.C.). The Emperor Taizong of Tang once said: "Agriculture is the basis of all things. People are the basis of a nation and food is the necessity of the people. The farming season is crucial to produce food and clothing for a nation."

The sowing and harvest of crops are decided by sun exposure and change according to different seasons. As early as the Spring and Autumn Period and Warring States Period, ancient Chinese people had determined the 4 earliest Solar Terms as the Summer Solstice, Spring Equinox, Autumn Equinox and Winter Solstice. They also realised that on the Summer Solstice, the daytime is the longest with the shortest night while

- 惊蛰花信：桃花

古代以每五日为一候，一年共七十二候。花信是指从第一年的小寒至第二年的谷雨的八个节气中，每一候中开花的植物中花期最准确的代表植物。桃花是惊蛰节气开始的第一候开花的植物。

Signal Flower of Awaking of Insects: Peach Blossom

In ancient times, Chinese people defined every five days as a pentad and there are totally 72 pentads in a year. The signal flowers are referred to the representative flowering plants, owning the most accurate and invariable flowering phase compared with others in each pentad of the 8 Solar Terms from the Minor Cold in the first year to the Grain Rain in the second year. The peach blossom is the signal flower of the first pentad of the Awaking of Insects.

on the Winter Solstice, people experience the longest night and shortest daytime, and on the Spring Equinox and Autumn Equinox, the lengths of daytime and night are the same. During the Qin and Han dynasties, in terms of the changing position of the sun on the orbit, one year time could be divided into 24 Solar Terms with respective name. In 104 B.C., *Taichu Calendar* was officially enacted by Emperor Wu of the Han Dynasty, which regulated the 24 Solar Terms. The *Qi* in the 24 Solar Terms is referred to the climate and the Solar Term means the beginning of a season. The 24 Solar Terms are: the Beginning of Spring, Rain Water, Awaking of Insects, Spring Equinox, Pure Brightness, Grain Rain, Beginning of Summer, Grain Buds, Grain in Ear, Summer Solstice, Minor Heat, Major Heat, Beginning of Autumn, End of Heat, White Dew, Autumnal Equinox, Cold Dew, Frost's Descent, Beginning of Winter, Minor Snow, Major Snow, Winter Solstice, Minor Cold and Major Cold.

The 24 Solar Terms are put in order according to spring, summer, autumn and winter, and in each term corresponding farming work should be arranged. This is also the reason why this

指气候，节气是指一个季节的开始。二十四节气分别为：立春、雨水、惊蛰、春分、清明、谷雨、立夏、小满、芒种、夏至、小暑、大暑、立秋、处暑、白露、秋分、寒露、霜降、立冬、小雪、大雪、冬至、小寒、大寒。

二十四节气按照春、夏、秋、冬的季节变化来排列，每个节气都有与之相对的农事安排，因此二十四节气是指导耕种、田间管理、秋收等农业生产的重要历法。在从事农业生产时，人们要以二十四节气为农业气候历，不违背农时（适宜耕种、收获的时节），耕种收获，应时劳作。

立春是一年中的第一个节气，是春季的开始，虽然寒冷的冬天还没有结束，但是天气逐渐开始变暖，河水逐渐解冻。俗语说"立春一年端，种地早盘算。"立春开始，农民就要为耕种做好准备。雨水节气，随着气温的慢慢回升，降水也逐渐增加，在春雨的滋润下万物开始复苏。农谚说"雨水有雨庄稼好，大春小春一片宝。"雨水这天如果降雨就预示着庄稼能够有一

calendar is important to guide people to manage the cultivation, crop growth and autumn harvest. In terms of agricultural production, people should treat the 24 Solar Terms as the agricultural schedule, without violating the proper farming season (the season that is suitable for cultivation and harvest), in order to plant and reap crops in season.

The Beginning of Spring is the first Solar Term in a year, regarded as the start of the spring. Yet a complete end is not put to the coldness of winter, the weather is becoming increasingly warmer and lakes begin to thaw. A common saying is stated as: "Farming work should be planned in the Beginning of Spring as the start of a year." From the Beginning of Spring, Chinese peasants are well prepared for the cultivation. For the Rain Water, with the slowly rising temperature, the rainfall increases gradually. With its moisture, all plants return to life. A farmer's saying as quoted "if it rains on the Rain Water, there will be a great harvest", indicates that a rainfall on the Rain Water will bring a good harvest that year. Since the temperature in the Awaking of Insects is continuously increasing, the soil defrosts and occasionally the spring thunders

• 春耕
Spring Ploughing

个好收成。惊蛰因气温不断上升，土壤解冻，不时有春雷打响，地下冬眠的动物都被惊醒开始活动而得名。伴随着惊蛰，春耕播种开始。春分这一天太阳直射地球赤道，全球昼夜等分。此时，中国大部分地区的越冬作物进入春季生长阶段，需要加强田间管理，要及时补充肥

are heard, which wakes up the dormiant animals and impels them to get outside. This is also the reason why this term is called as the Awaking of Insects. Along with the Awaking of Insects, the spring ploughing and sowing begin as well. On the Spring Equinox, the sunlight shines vertically on the equator, equally dividing the day and night. At that time, the winter crops of the majority of regions in China enter the spring growth stage and the field management should be strengthened with timely supplement of fertilizer. A famous farmer's saying prevailing in central China: "The wheat sprouts on the Spring Equinox and every minute counts." The Pure Brightness derives from the southeast monsoon (pure and bright wind), when plants bud and weather becomes warmer and brighter. With regards to the farm works in this period, it is said "melons and peas should be cultivated around the Pure Brightness". Not limited to one of the 24 Solar Terms, the Pure Brightness is also a traditional festival in China, on which people will sweep tombs, offer sacrifices to ancestors and go out for a spring

料。中部地区有一句农谚："春分麦起身，一刻值千金。"清明时节草木萌动，天气逐渐变得温暖、清澈，故名。在农事安排上，有"清明前后，种瓜种豆"的俗语。清明不仅是二十四节气之一，也是中国传统节日，这一天人们要扫墓祭祖、外出踏春。谷雨是春季最后一个节气，古人有"雨生百谷"的说法，因此得名。此时的降雨对五谷的生长十分有利。

立夏是夏季的开始，此时农作物进入旺盛生长的阶段，田地里的杂草生长也较快，必须勤于铲除杂草，古人有"立夏三朝遍地锄"的说法。同时，随着温度的升高，还要做好防旱的准备。小满时节，夏熟的农作物已经结果，但又尚未全熟，进入了成长的最后阶段，此时要保证农作物生长所需的水分。芒种节气是麦类等有芒农作物已经成熟开始收获、夏种作物开始播种以及加强春播作物田间管理的时期，是一年中农民最繁忙的季节。夏至这一天，太阳直射北回归线，北半球地区的白昼最长、黑夜最短。从夏至至立秋，是一年中最热的一段

hiking. The Grain Rain is the last term of spring and its name is based on the old saying of ancient Chinese people "rain breeds cereals". The rain during the Grain Rain is beneficial for the five cereals' growth.

Summer starts with the Beginning of Summer and all crops enter the period of vigorous growth, along with the rampant weeds on the field. It is said that: "The weeds should be widely hoed three days after the Beginning of Summer." This proverb reminds people to treat and uproot weeds diligently. At the same time, with the increasing temperature, peasants need to prepare for the drought. The crops ripening in summer step into the last growing phase which yield fruits but not fully ripen in the term of the Grain Buds. Therefore, it is important to ensure the water supply for the crops. In the Grain in Ear, the crops with awn like the wheat are mature and need to be reaped. People also start to sow the seeds of the summer crops and strengthen the field management of spring plants, known as the busiest period for peasants all the year round. On the Summer Solstice, the sunlight shines vertically on the Tropic of Cancer and the Northern Hemisphere experiences the longest daytime and

● 小暑节气石雕（图片提供：FOTOE）
Stone Carving of Minor Heat

时期，也是农作物快速生长的季节。小暑是进入夏季后的第三个节气，"小"代表炎热的程度，此后天气会越来越炎热。大暑则标志着进入一年中最炎热的季节，此时太阳光线强烈、雨水充沛，是各种农作物生长最快的阶段。在这一阶段，人们要做好防汛防旱、防暑降温工作。

立秋是秋季的开始。秋季是庄稼成熟、收获的季节。立秋的到来表明农作物的秋收即将开始。此

shortest night. The period between the Summer Solstice and the Beginning of Autumn is not only the hottest in a year, but also the time for crops to rapidly grow. The Minor Heat is the third term after entering summer and its name indicates the level of hotness, after which the weather will be further hotter. As the hottest period of a year, the Major Heat brings strong sunshine and abundant rainfall, providing excellent conditions for plants to maintain the fastest growth rate. It is crucial to be well prepared for the drought, heatstroke prevention and

后，气温开始下降，昼夜温差逐渐加大，农谚有"早上立了秋，晚上凉飕飕"的说法。"处"，就是"去"，意思是过去、结束。处暑就是炎热夏季的结束，表明暑气至此就要终止了。白露节气，气温开始下降，天气转凉，早晨的草木上开始凝结露水。此时，水稻逐步进入灌浆成熟期，这时如果秋高气爽、光照充足，对灌浆、晾田都极为有利，所以农谚有"白露天晴稻象山"的说法。秋分这一天，太阳又一次直射地球赤道，昼夜等长，此后太阳直射点逐渐南移，北半球变得昼短夜长，气温也越来越低。寒露时节，气温下降，露水也开始有了寒意。农谚说："寒露一到百草枯。"这意味着温度已经达到农作物生长发育的下限。寒露是秋收、秋种的关键季节，人们要注意秋旱和秋季连阴雨等不利天气对秋收、秋种的影响。寒露之后就到了霜降节气，气温越来越低，这时要预防农作物受到冻害影响。此时正处于秋冬过渡期，气温变化幅度较大，人们要抓紧有利的天气，搞好秋收、秋种。

cooling.

Autumn is a season for maturity and harvest. As the start of autumn, the Beginning of Autumn manifests that the autumn harvest is coming. From now on, the weather begins to cool and daily temperature difference widens, as the farmer's saying states: "The Beginning of Autumn comes in the morning and cool arrives in the evening." *Chu,* known as passing away in Chinese, is referred to the past and the end. The burning hot summer finishes at the End of Heat, marking that the summer heat will pass away from this term. The arrival of the White Dew indicates a dropping temperature and cooling weather. In the morning, the dew can be seen on the plants. In this term, the paddy rice turns into the filling and ripening stage and the air and climate is clear and crisp with adequate sunshine are extremely beneficial conditions for filling and field drying, in accordance with the saying:"If it clears up on the White Dew, there will be mountains of paddy rice." On the Autumn Equinox, the sunlight shines vertically on the equator again, making the equal lengths of daytime and night, and moves southward with short day and long night in the Northern Hemisphere.

● 《耕图·收刈》（清）
Painting of Farming and Reaping (Qing Dynasty, 1616-1911)

立冬是冬季的开始，人们要加强三麦（小麦、大麦、元麦）和油菜的冬前管理，保证壮苗越冬。立冬之后是小雪节气，气温已经降到零度以下，开始出现降雪，但雪量不大，人们要加强越冬作物的田间管理，保证秋种作物的生长。大雪节气的雪量要比小雪节气大，中国有"瑞雪兆丰年"的俗语，覆盖大地的积雪，能冻死害虫，为越冬

The temperature drops step by step as well. Because of the further decreasing temperature, the dew presents the chill in the term of Cold Dew. It is prevailing that "plants wither when the Cold Dew comes", means that the temperature drops to the level that will prevent the crops from growing. The Cold Dew is critical period for autumn harvesting and sowing and people pay attention to the adverse weather like the seasonal drought and continuous rain that pose threats to the farming work. After the Cold Dew, the arrival of Frost's Descent leads to the decreasing temperature and the possibility of freezing injury should be noticed and avoided. This is also the transitional period of autumn and winter when the temperature drops heavily. It is suggested to firmly grasp the opportunity of finishing a good job of harvesting and sowing.

Winter starts with the Beginning of Winter. People in this term should

作物创造良好的越冬环境。冬至，太阳直射南回归线，北半球地区的白昼最短，黑夜最长，也就意味着进入一年中最寒冷的阶段。按照民间习惯，从冬至的第二天开始数九（每九天算一九），淮北地区有一首冬至数九的《九九歌》："一九二九不出手，三九四九冰上走，五九六九沿河看柳，七九河开，八九雁来，九九加一九，耕牛遍地走。"这首民谚形象地描述

● 饺子
Dumplings

enhance the management of the three cereals (wheat, barley and highland barley) and rape, in order to ensure the seedlings to be strong enough to live through the winter. After the Beginning of Winter, the temperature drops below zero degree and snow of limited volume begins to fall in the Minor Snow. The field management of the overwintering crops should be emphasized to make sure the growth of autumn crops. The snowfall of the Major Snow is heavier than that of the Minor Snow and Chinese people believe in the proverb as: "A timely heavy snow promises a good harvest." The snow covering on the earth can destroy the harmful insects, which is beneficial to overwintering crops. On the Winter Solstice, the sunlight shines vertically on the Tropic of Capricorn, with the shortest daytime and longest night in the Northern Hemisphere, namely the coldest period of a year. According to folk customs, Chinese people count the nine-day as a period from the 2nd day after the Winter Solstice (every nine days is one nine-day period), and the famous *Nine-day Ballad* for counting the nine-day period prevailing in the northern region of Huaihe River is generally shown as follows: "For the

- **大寒时节的雪乡**

  雪乡位于黑龙江省牡丹江市,从每年的10月至次年5月降雪长达7个月,积雪厚,景色美,是著名的"中国雪乡"。

  ### Snow Country on the Major Cold

  The Snow Country is located in Mudanjiang City of Heilongjiang Province and from October to May next year, it snows for 7 months here with thick snow cover and beautiful scenery, known as the famous Snow Country in China.

了冬至以后八十一天左右的气候状况。小寒节气时值三九前后，俗话说"热在三伏，冷在三九"，预示着这时天气更加寒冷。大寒，是天气寒冷到极点的意思，是二十四节气中的最后一个节气，人们要注意越冬作物的防寒防冻，以保证第二年的收获。

first and second nine-day periods, we dare not hold out our hands; for the third and fourth, we walk on the frozen path; for the fifth and sixth, we admire willows along the river; for the seventh, the river thaws; for the eighth, wild geese return and winter finally draws to an end in the last nine-day as cattle walk outside the barn." This ballad vividly reflects the changing weather of approximately 81 days after the Winter Solstice. The Minor Cold is around the third nine-day period, which illustrates the folk saying "hot in between the Minor Heat and Major Heat and cold in the third nine-day period", indicating that the weather will become colder afterwards. It goes extremely cold after entering the Major Cold, known as the last one of the 24 Solar Terms, in which people pay close attention to the measures of cold-and-freeze-proofing for the overwintering crops, in order to ensure the harvest in the next summer.

> 农学家与著作

在中国古代"重农"思想的影响下,农业得到了广泛的发展,同时还出现了许多优秀的农学家以及集合了当时最为先进的农学技术的农学专著。

贾思勰与《齐民要术》

贾思勰(生卒年不详)是北魏时期杰出的农学家。他出身农民家庭,从小就很喜欢读书,尤其重视对农业生产技术知识的学习和研究。他撰写了涵盖农、林、牧、副、渔等方面的农学专著《齐民要术》。书名中的"齐民",指普通百姓,"要术"指谋生方法。

《齐民要术》分为10卷,共92篇,11万字左右。第一卷记述耕田、收种、种谷;第二卷记述了谷

> Agriculturists and Works on Agronomy

In ancient China, influenced by the thought of valuing agriculture, the agriculture was widely developed. There were many outstanding agriculturists as well as numerous famous books of agriculture collecting the most advanced agricultural techniques at that time.

Jia Sixie and *Main Techniques for the Welfare of the People*

Jia Sixie (unknown years of birth and death) is an outstanding agriculturist of the Northern Wei Dynasty (386-534). He was born in a farmers' family and loved reading from childhood, especially concentrating on the study and research on agricultural techniques and knowledge. He wrote the famous book *Main Techniques for the Welfare of the People (Qi Min Yao Shu)*, covering the areas of agriculture, forestry, animal

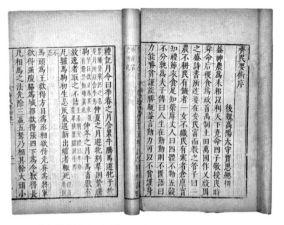

《齐民要术》书影（明刻本）
Photocopy of *Main Techniques for the Welfare of the People* (Ming Dynasty, 1368-1644, Block-printed edition)

类、豆、麦、麻、稻、瓜、芋等；第三卷记述了葵菜、蔓菁等的种植方法；第四卷记述了枣、桃、李等果树的栽培方法；第五卷记述如何种桑养蚕，榆树、白杨、竹以及染料作物的栽培方法、伐木技术；第六卷记述了畜、禽及鱼的养殖；第七卷记述货殖、酿酒等；第八、第九卷记述了酱、醋、乳酪的酿造及储存，煮胶、制墨等；第十卷记述了北魏域外的物产，包括100多种热带、亚热带植物和60多种野生可食用的植物。该书内容丰富，涉及面极广，包括农作物的栽培、经济

husbandry, side-line production and fishery. With regards to the book name, *Qi Min* means ordinary people and *Yao Shu* refers to the way to make a living.

Main Techniques for the Welfare of the People is consisted of 10 chapters of approximately 110,000 words, including totally 92 articles. The first chapter is about ploughing, seed-collection and sowing; the second chapter is about cereal, bean, wheat, flax, paddy rice, melon and taro; the third chapter is about the cultivation of curled mallow and turnip; for the fourth chapter, the planting of fruit trees like the jujube, peach and plum trees is recorded; how to cultivate mulberry, silkworms, elm, white poplar and bamboo and to plant dye crops as well as lumbering are included in the fifth chapter; in the chapter six, the breeding of domestic animals, fowls and fishes is described; the seventh chapter is about commercial activity and wine-making; in the eighth and ninth chapters, the production and storage of sauce, vinegar and cheese, as well as gum-boiling and ink-making are involved; for the final chapter, it introduces the natural products outside

林木的生产以及各种野生植物的利用等。同时书中还详细介绍了各种家禽、家畜以及鱼、蚕等的饲养和疾病防治，并把农副产品、食品加工、日用品生产等内容都包含在内。在《养羊篇》中，贾思勰指出公羊与母羊的比例应为1:5，如果公羊太少，母羊很难受孕繁衍后代；如果公羊太多，则会造成羊群纷乱。书中还列举了形式多样的耕作方式，如深耕、浅耕、初耕、转耕、纵耕、横耕、顺耕、逆耕以及春耕、夏耕、秋耕、冬耕等，并详细说明了每一种耕作方式适用于哪

the Northern Wei, including more than 100 tropical and subtropical plants and more than 60 edible wild plants are written about. The book contains various contents and involves an extremely wide range, covering the cultivation of crops, production of trees with economic values and the utilisation of different wild plants. At the same time, it also introduces the details of the feeding, prevention and cure of diseases of various fowls, domestic animals, fishes and silkworms. In addition, the subsidiary agricultural products, food processing and the production of daily necessities are touched as well. In the article named as *Sheep Raising*, Jia Sixie pointed out the number of ewes should be five times of that of rams and according to this proportion, he concluded that if the quantity of rams is limited, it is difficult for the ewes to reproduce; if the ram is too many, the disorder will occur in the sheep flock. Meanwhile, various forms of ploughing methods are listed in the book as well, such as deep ploughing, shallow ploughing, first ploughing, transferred ploughing, vertical ploughing, horizontal ploughing, ordinal ploughing and reverse ploughing, as well as spring ploughing, summer ploughing, autumn ploughing

● 铁犁（汉）
铁犁出现于战国时期，是古代农业生产中的耕翻农具。
Iron Plough (Han Dynasty, 206B.C.-220A.D.)
The iron plough, emerged in the Warring States Period (475B.C.-221B.C.), is a traditional ploughing tool in ancient agricultural production.

些情况、如何操作等实际内容。

《齐民要术》是一部农业科学技术巨著，记录了大量古代农业生产的宝贵经验，推动了古代农业生产的发展，被誉为古代的"农业百科全书"。

王祯与《王祯农书》

王祯是元代著名农学家，他十分重视农业生产，认为吃饭是百姓的头等大事。王祯在做地方官期间，主要的政绩就是抓农业生产。他亲自指导民众耕织、种植、养畜等。他留心农事，处处观察，积累了丰富的农业知识，这为其撰写《王祯农书》提供了丰富的材料。

《王祯农书》约有13.6万字，配有281幅插图，共分37卷，包括《农桑通诀》6卷、《百谷谱》11卷、《农器图谱》20卷。在《农桑通诀》中，王祯专辟灌溉篇，引用了各种例证说明兴修水利是中国自古以来的优良传统。他还介绍了多种引水灌溉的方法，把农田灌溉摆在重要地位，并指出南方兴修水利、除水害的具体途径。在《百谷谱》中，他对谷属、蔬属、果属、竹木

and winter ploughing with concrete descriptions of the applicable situations and practical operational process.

Main Techniques for the Welfare of the People is a masterpiece of agricultural techniques, recording large amounts of precious experience on ancient agricultural production and promotes Chinese agriculture to develop. It enjoys the reputation of the ancient Encyclopedia of Agriculture.

Wang Zhen and *Wang Zhen's Book of Agronomy*

Wang Zhen was a famous agronomy expert in the Yuan Dynasty (1206-1368). According to him, agricultural production was the first priority and food was of the first importance in people's life. As a local official, his main job was to promote agricultural production. He taught local farmers weaving, planting and livestock keeping, and studied from other's experience by close observation and repeated practice. His rich agricultural knowledge and materials have greatly helped him with his book.

Wang Zhen's Book of Agronomy has 136,000 characters and 281 illustrations. This book is divided into 37 volumes: 6 volumes of *Rules of Agriculture and Silk*

● 水排模型（图片提供：微图）

此水排按照元代《王祯农书》中的立轮水排按1：10的比例制作而成。

Model of Water Wheel

This model was made on a scale of 1 to 10 based on the illustration of standing water wheel in *Wang Zhen's Book of Agronomy* in the Yuan Dynasty (1206-1368).

及其他杂类植物的性状进行了详细描述。另外，书中还总结了马、牛、羊、猪、鸡、鸭、鹅等牲畜、家禽的圈养方法，总结了养蚕的一些新经验，指出在择种收种、保存蚕种、饲养管理、调节室温等方面的注意事项。

王祯不仅是著名的农学家，还是一位出色的农业机械学家，他创制了很多农具。例如，他创制了适应南方水田土壤的有六齿或四齿的

Farming, 11 volumes of *Chart of One Hundred Cereals*, 20 volumes of *Illustrations of Agricultural Tools*. In the *Rules of Agriculture and Silk Farming*, there is one chapter on irrigation where Wang Zhen listed many examples to prove that making use of irrigation and water conservancy projects was historically the Chinese farmers' tradition. He also described different types of irrigation and pointed out the significance of developing water conservancy projects in southern

- **耧车模型**（图片提供：微图）

此耧车根据元代《王祯农书》等资料制作而成。耧车是条播机械，由一人在前牵牛拉耧，一人在后挽耧进行播种，在当时是一种先进的农具。

Model of Seed Plough

This model was made based on historical materials including *Wang Zhen's Book of Agronomy* in the Yuan Dynasty (1206-1368). Seed plough has adopted the concept of seed drill that was operated by two people: one in the front took the lead, followed by one person operating the plough to seed the field. It was then considered a very advanced agricultural tool.

耕垦工具铁塔，能在泥中行走、便于水田作业的工具秧马，适用于水田中除草的耕耘工具耘荡和适用于华北平原畜力耕耘的器具耧锄等。

宋应星与《天工开物》

宋应星（1587—？），明代科学家。他对中国各地的农业和手工业做了广泛的社会调查，虚心向当地的群众请教，收集了丰富的资料，所撰写的《天工开物》于1637年刊行。

《天工开物》内容丰富，不仅包括农业，还对手工业、工业等方

China in order to prevent any future high water damage. In the *Chart of One Hundred Agricultural Products*, he made detailed descriptions of many plants including grain species, vegetable species, fruit species, bamboo and wood species. He also summarized his experience in captive breeding of horse, ox, ram, pig, rooster, duck and goose, as well as new techniques of silkworm breeding and notes on egg selection and collection, egg keeping, breeding management, temperature adjustment, etc.

More than an agriculturist, Wang Zhen was also an expert in agricultural machinery. He has invented many agricultural tools, including the farm rake with four or six teeth designed exclusively for farmlands in southern China, boat-shaped seedling machine to

面进行了系统的总结，形成了一个完整的科学技术体系。英国达尔文称此书为"权威著作"，李约瑟则称此书为"17世纪早期的重要工业技术著作"。

《天工开物》详细叙述了各种农作物和工业原料的种类、产地、生产技术和工艺装备以及生产经验，既有大量确切的数据，又配有123幅精美的插图。全书共分为上、中、下三卷。上卷记载了谷物、豆、麻的栽培和加工方法，蚕丝的纺织和染色技术以及制盐、制糖工艺。中卷记载了砖瓦、陶瓷的制作，车船的建造，金属的铸锻，煤炭、石灰、硫黄和白矾的开采和烧制以及榨油、造纸的方法。下卷记载了金属矿物的开采和冶炼，兵器的制造，颜料、制酒以及珠玉的采集与加工。

徐光启与《农政全书》

徐光启（1562—1633），明代杰出的科学家。他因结识罗马的传教士利玛窦，跟随他学习了很多西方自然科学知识。《农政全书》是徐光启一生最杰出的作品，他与西

help farmers work in the mud fields, the weeding hoe used in paddy fields and the animal-drawn seed plough applied in North China Plain, etc.

Song Yingxing and *Exploitation of the Works of Nature*

Song Yingxing (1587-?), a scientist in the late Ming Dynasty (1368-1644). He conducted a social survey on the agricultural and handicraft industries in many places of China studied from local people's experience and gathered rich troves of data. *Exploitation of the Works of Nature* was presumed in 1637.

The *Exploitation of the Works of Nature* not only covers the agricultural industry, but also gives a systematic summary of the handicraft industry as well as other industries. It overall constructs a complete framework of scientific and technological development in China at that time. It was considered "an authoritative work" by Charles Darwin and claimed as "a great work of industrial technologies in the early 17th century" by Joseph Needham.

The *Exploitation of the Works of Nature* gives a deliberate description of all field crops and the type, place of origin, production technology and craft

- **《耕田图》**

中国古代是一个以农业为主的社会，殷商时期的甲骨文中就已经出现了"犁"字。春秋战国时期，牛耕已得到初步使用。

Picture of Ploughing Work

The ancient Chinese society was an agriculture-based society. Early in the Shang Dynasty (1600 B.C.-1046 B.C.), people used the pictographic script of the Chinese character 犁 (*Li*, plough). The ox-driven plough was firstly used in the Spring and Autumn Period (770 B.C.-476 B.C.) and the Warring States Period (475 B.C.-221 B.C.).

equipment of industrial raw materials as well as production experience, all supported by a huge amount of statistics and 123 evidence-based illustrations. This book is divided into three volumes: the first volume includes cultivating and processing cereals, beans and fiber plants, silk textile and dyeing technologies, and salt and sugar technologies; the second volume includes brick, tile and porcelain manufacturing process, ships and carts, metallurgy, exploitation and firing technologies of coal, lime, sulfur and alum, oil technology and papermaking technology; the third volume includes metallurgy of metallic minerals, weapon production, pigments, fermented beverages and the collection and processing of pearls and jade.

Xu Guangqi and *Pandect of Agronomy*

Xu Guangqi (1562-1633) was an outstanding scientist in the Ming Dynasty (1368-1644). Then he made acquaintance with Matteo Ricci, the Roman missionary, who taught him the western system of natural science. *Pandect of Agronomy* was his greatest contribution and brought him the title of one of the Four Great Agriculturists in Ancient China. The other three agriculturists are: Fan

汉的氾胜之、北魏的贾思勰、元代的王祯并称为"中国古代四大农学家"。

《农政全书》共60卷，70多万字，全书共分为农本、田制、农事、水利、农器、树艺、蚕桑、蚕桑广类、种植、牧养、制造、荒政"十二门"。在《树艺》一章中，徐光启详细记述了番薯的种植、贮藏和加工法，并提到番薯育苗越冬、剪茎分种、扦插、窖藏干藏等技术。《农政全书》是中国最早系统介绍番薯种植方法的著作。此书篇幅比《齐民要术》多约7倍，是中国古代农书中篇幅最多的一部，也是中国第一部综合性农学著作。

Shengzhi in the Western Han Dynasty (206 B.C.-25 A.D.), Jia Sixie in the Northern Wei Dynasty (386-534), and Wang Zhen in the Yuan Dynasty (1206-1368).

Pandect of Agronomy has sixty volumes including more than 700,000 characters, divided into twelve parts: the fundamentals of agronomy, field system, farming tasks, irrigation, agricultural tools, horticulture, sericulture, types of silkworm, forestry, animal husbandry, manufacturing, and land reclamation. In the *Horticulture* part, Xu Guangqi has kept a detailed account of the cultivation, storage and process of sweet potatoes, along with some technologies including planting sweet potatoes in wintertime, stem cutting and propagation, cuttage, cellar storage of sweet potatoes, storage of dried sweet potatoes, etc, which is the earliest systematic introduction about sweet potato. The length of *Pandect of Agronomy* is about seven times the length of *Main Techniques for the Welfare of the People*, and it is the longest book among all Chinese ancient agricultural works. It is also the first comprehensive Chinese treatises on agriculture ever.

- 徐光启像
  Portrait of Xu Guangqi

> 水利灌溉工程

中国古代十分注重农业生产，为了保证和促进农业生产，古人建造了许多卓越的水利工程来灌溉农田，并开创了挖渠引水、凿井汲水等先进方法。

都江堰

都江堰位于四川省成都平原西部的岷江上。秦昭襄王五十一年（前256年），秦国蜀郡太守李冰父子奉命修建。开始时称作"湔堋"，到宋代，才正式称作"都江堰"。"堰"就是挡水的堤坝。

当时，岷江下游泥沙堆积，雨季时洪水泛滥，而其他季节则十分干旱缺水，对岷江两岸的农业生产影响极大。李冰父子吸取前人的治水经验，采用"深淘滩，低作

> Water Conservation and Irrigation Projects

Ancient Chinese significantly valued agriculture, thus they built many great water conservation and irrigation projects to promote agricultural production. They also invented many advanced technologies regarding to digging canals and wells.

Dujiangyan Irrigation Project

Dujiangyan Irrgation Project is located in the Minjiang River in Chengdu Plain, Sichuan Province. In 256 B.C., the local governor Li Bing and his son were put in charge of this project, which was originally called *Jianpeng* (watershed dike) and was later renamed into *Dujiangyan* in the Song Dynasty (960-1279). *Yan* means dam in ancient Chinese.

At that time, the downstream riverbed of Minjiang River was filled

- 宝瓶口
  Precious-bottle-neck

堰"的方法，率领当地民众一同治理岷江。

李冰父子所采用的"低作堰"的方法，是在岷江峡谷内用石块砌成石埂，因其形状似鱼头，故名"鱼嘴"。"鱼嘴"将岷江分为供灌溉的内江和供岷江主流泄洪的外江。内外江之间有一座"宝瓶口"，具有节制水流的作用。在"鱼嘴"的下方还有一段溢洪排沙的河道，名为"飞沙堰"。鱼嘴和宝瓶口联合运用，能按照灌溉、防

with silt. The river was plagued by flooding in the rainy season and droughts at other times, which has severely affected the agricultural production along riverside areas. Based on their predecessors' experience, Li Bing and his son adopted the principle of Digging the Riverbed Deeper and Building Weirs Rather Than Big Dams and organized local people to working on this project

Li Bing and his son built a levee with rocks in the valley of Minjiang River, which was named the Fish Mouth Levee

洪的需要分配洪、枯水流量。李冰父子还采用"深淘滩"的方法，使河床保持一定的深度，保证在发生洪灾时大量水流能够安全通过。

都江堰变害为利，让成都平原成了"天府之国"。它是世界上修筑年代最久且仍被使用的水利工程，被誉为"世界水利文化的鼻祖"。2000年，联合国世界遗产委员会第24届大会上，都江堰水利工程因其历史悠久、规模宏大、布局合理、运行科学且与环境和谐结

as its conical head was said to resemble the mouth of a fish. This levee divides the Minjiang River into the inner stream that is exclusively for irrigation and the outer stream that is the main stream for flood discharge. Between these two steams is the Bottle-neck Channel, gouged by Li Bing to better distribute the water streams. Down the Fish Mouth Levee stands the Flying Sand Weir for draining silt. The levee and weir work together to distribute water streams according to the agricultural irrigation demand in both rainy and dry seasons, and at the same time prevent flooding by letting water smoothly pass every section of the river even in flood season.

Dujiangyan Irrigation Project has turned the Chengdu Plain into the most productive land in China. It is the oldest existing water conservation and irrigation project and is still in use today. In 2000, at the 24th conference of the UNESCO World Heritage Committee, Dujiangyan

- 都江堰石人
  李冰还在进水口放了三个石人，使水"竭不至足，盛不没肩"。这些石人是最原始的水尺。
  Stone Statues at Dujiangyan Irrigation Project
  Li Bing placed three stone statues on the upper reach of Dujiangyan's water intake to measure water level: the water should exceed their ankles in drought season and should not drown their shoulders in flood season. These statues were the earliest water gauge.

合，在历史和科学方面具有突出的价值，而被确定为世界文化遗产。

郑国渠

公元前246年，韩国国君派水利工程人员郑国去往秦国，游说秦国在泾水和洛水间修建一条大型灌溉渠道，表面上说是为发展秦国农业，而真实目的则是要消耗秦国的国力。秦王认为修建灌溉大渠乃是为秦国建造一座"天下粮仓"，于是便接受了郑国提出的建议，批准了灌溉渠道的修建。

郑国修建的这条渠是以泾水为水源，对渭水北面的农田进行灌溉的水利工程。郑国创造了"横绝"技术，就是使渠道跨过冶峪河、清河等大小河流，拦截水流使之进入渠中，从而增加水流量。同时，郑国还利用横向环流的离心力作用，让流入渠道的表层水流流向凹岸，而底部水流流向凸岸，巧妙地解决了粗沙入渠堵塞渠道的问题。

这条渠建成15年后（前221年），秦国消灭了其他六个诸侯国，建立了统一的王朝秦朝。嬴政感念郑国修渠有功，下令将此渠命

Irrigation Project was listed as a World Cultural Heritage site due to its grand history and size, smart layout and running rationale, its harmonious relationship with surrounding environment and outstanding historical and scientific values.

Zheng Guo Canal

In 246 B.C., King of the State Han sent its water conservation engineer Zheng Guo to the Qin State in order to convince the king of the Qin State to construct a large canal connecting the River Jing with the River Luo, in the name of developing the Qin State's agricultural industry yet out of a plan to drain the Qin State's energy and wealth. The king of Qin was convinced that this canal would greatly boost its agriculture and accepted Zheng Guo's suggestion to build this canal.

Zheng Guo used the water resource from the River Jing to irrigate farmlands north to the River Wei. He created the strategy of Cutting Rivers to Collect Water Resource, namely feeding the canal with as many water streams as possible by cutting major and small rivers (including Yeyu River and Qing River) and channeling their water streams into the canal. In addition, he smartly used the centrifugal force of transverse circulation

大禹治水

　　四千多年前，黄河流域频发水患。尧帝命鲧负责治水。鲧采取"水来土掩"的策略，治水失败，被尧处死。鲧的儿子大禹接替他主持治水。大禹吸取了父亲的教训，首先带着尺、绳等工具对当地的主要山脉、河流进行了实地考察和测量。他发现黄河之所以频发水患是因为河床淤积、流水不畅。于是，大禹一改父亲"堵"的治水方法，用"治水须顺水性，水性就下，导之入海。高处就凿通，低处就疏导"的方法，疏通了黄河河道，使洪水能够快速通过，消除了水患。

Yu the Great and Flood Control Project

More than four thousand years, people living by the Yellow River suffered greatly from frequent floods. Chief Gun was authorized by King Yao with developing a method to control the flood. Gun built many dikes to block the water from overflowing the river banks, however, his plan failed and he was executed by the king. Then the king put his son, Yu the Great, in charge of the project. Keeping his father's lesson in mind, Yu the Great firstly

• 大禹治水
Yu the Great Working on His Flood Control Project

conducted a careful study of local mountain ranges and water systems with his measuring tools, rope and ruler. He then discovered the reason of the frequent floods: the silted river bank has blocked the water stream. Therefore, he did not follow his father's strategy; instead he made use of the characteristics of the Yellow River and decided to channel it into the sea by tunneling through the high-rising sections of the river and dredging the low-lying sections. As a result, water traveled smoothly into the sea and the flood problem was successfully solved.

名为"郑国渠",这是中国历史上第一个以人名命名的工程。

龙首渠

元朔年间（前128—前123），汉武帝下令修建龙首渠,引洛河之水灌溉农田。龙首渠位于陕西蒲城、大荔一带。由于这一带土质疏松,不能用一般的修建方法,于是智慧的劳动人民发明了"井渠法",即在山坡上每隔300米打一眼竖井,使渠道从地下穿过,开创了隧洞竖井施工法的先河。至今新疆地区仍在使用这种井、渠结合的办法修建灌溉渠道,当地叫做"坎儿井"。而生活在中亚和西南亚干旱地带的人们也是用这种办法灌溉农田的。

of river steams: the upper water flows into the canal's water intake on the concave bank while the grits carried by the lower layer of water stream on the convex bank. Thus silting is never an issue for this canal.

In the fifteenth year after the construction of Zheng Guo Canal (in 221 B.C.), the Qin State successfully conquered the other six vassal states and unified China. Ying Zheng, the king of the Qin State, appreciated Zheng Guo's work and gave an order to name this canal as Zheng Guo Canal, which was the first national project named after a person's name in the Chinese history.

Longshou Canal

The construction of Longshou Canal lasted from128 B.C. to 123 B.C. under the order of Emperor Wu of the Han

● 坎儿井地下水渠（图片提供：微图）
Underground Canal of Karez

龙首渠的建成，使一万余公顷的盐碱地得到灌溉，粮食产量增加了好几倍。这段穿过商颜山（今陕西铁镰山）的地下渠道长达5公里，是中国历史上第一条地下井渠，在世界水利史上也是一个伟大的创造。

Dynasty, aiming to irrigate farmlands with water from the River Luo. Longshou Canal was located in Pucheng County and Dali County of Shaanxi Province. Those areas failed to use ordinary canal constructions due to their loose soil. Therefore local people came up with the innovative well-canal digging method, which is to dig wells at an interval of 300 meters and then dig an underground canal that connects those wells. This was the earliest practice of underground tunnel shaft construction. Today, canal constructions in Xinjiang are still following this method, which are called karez by local people. Additionally, people in some arid areas in central and southwestern Asia also used the karez for farmland irrigation.

The construction of Longshou Canal helped to irrigate the salty land over an area of more than 10,000 hectares, with some times increase of grain output. This underground canal is over 5 kilometers long, extending through Mount Shangyan (today's Mount Tielian in Shaanxi Province). It is the first underground canal in China as well as a great construction in the world's history of water conservation and irrigation projects.